W]

Cover photograph by Ian Wilson
A majestic bull elk bugling in the autumn morning mist, high in the Rocky Mountains of Canada

WILD & FREE
LIVING WITH WILDLIFE IN CANADA'S NORTH

Text and Photographs by
IAN AND SALLY WILSON
Illustrations by Sally Tatlow Wilson

Gordon Soules Book Publishers Ltd.
West Vancouver, Canada
Seattle, U.S.A.

**With special thanks to Maureen Colclough
for all her help with this book**

First Printing September, 1989
Second Printing October, 1989

Text © 1989 by Ian Wilson
Illustrations © 1989 by Sally Tatlow Wilson
All rights reserved. No part of this book may be reproduced in any form by any means without the written permission of the publisher, except by a reviewer, who may quote passages in a review.

Published in Canada by
Gordon Soules Book Publishers Ltd.
1352-B Marine Drive
West Vancouver, B.C.
Canada V7T 1B5

Published in the U.S.A. by
Gordon Soules Book Publishers Ltd.
620 - 1916 Pike Place
Seattle, WA 98101

Canadian Cataloguing in Publishing Data

Wilson, Ian, 1955–
Wild & free

Includes bibliographical references.
ISBN 0-919574-87-4

1. Zoology - Canada, Northern. 2. Canada, Northern - Description and travel. I. Wilson, Sally Tatlow, 1955–   II. Title.
QL221.N67W54   1989   591.971   C89-091441-9

Thanks to Murray O'Neill for the photo on page 67
Printed and bound in Canada by Hignell Printing Limited

CONTENTS

	Preface	7
1	It's a Wild Life	9
2	Bugler of the Forest	11
3	Porky	21
4	Black Bear Ahead!	30
5	Woodland Drummer	38
6	Curious Caribou	45
7	Little Haymaker	53
8	Winged Hunters	58
9	Goat Country	81
10	Call of the Wild	90
11	The Sassy Squirrel	98
12	Wedge of Wings	105
13	Mule Deer	114
14	Alpine Whistler	121
15	Feathered Friends	129
16	The Pine Marten	136
17	King of the Mountain	145
18	Smaller Friends	153
19	Busy Beavers	161
20	Calling All Moose	170
21	Through the Viewfinder	179
22	A Wilderness Legacy	183
	References	187
	Index	188

Authors Ian and Sally Wilson

PREFACE

The adventures that we recount in this book had their beginnings several years ago when Sally and I met while ski-mountaineering near Whistler, British Columbia. I was sharing a sandwich with a grey jay when Sally joined us for lunch.

After chatting awhile, Sally and I discovered that we shared a love for nature and the outdoors. We both craved the freedom of wide open spaces, the thrill of following goat trails across mountain tops, and the feeling of solitude on lonely lakes. One thing led to another, and before long we came up with the crazy idea of making a break from our city jobs and living in the wilderness of northern Canada.

That wilderness sojourn lasted fourteen months and inspired us to spend as much time as possible in the north. It also changed our lives. We decided to become full-time writers and photographers. Our first book, *Wilderness Seasons: Life and Adventure in Canada's North*, recounts our experiences through those fourteen months.

Many readers of *Wilderness Seasons* wrote to tell us they would enjoy reading more about the animals we encountered during our travels. We took up that challenge and spent many seasons living with wildlife in the north. This book is the result. Each chapter is a blend of our personal experiences and interesting facts about a bird or animal. We invite you to come with us down the winding trail of our adventures and misadventures with wildlife.

Ian and Sally Wilson
July, 1989

TOP: *Feeding a porcupine—carefully!*
BOTTOM: *Getting close to a mountain sheep*

ONE

It's a Wild Life

"If you're going hunting with only a camera in that country, remember three things: Black bears climb trees, moose don't, and grizzlies are not supposed to!"

With that cheerful bit of wisdom from an old-timer, Sally and I set forth on the first of many extended trips to photograph wildlife in northern Canada.

If only the old-timer had offered some advice about what to do when we encountered one of those animals. Within days of being flown into remote wilderness, we were startled by a grizzly who burst through the willows into the clearing where we were camping. The bear was enormous! We stood still, hardly daring to breathe, then backed away cautiously. I sincerely hoped the animal wouldn't take offense.

The bear stood up on its hind legs, sniffing the air and slowly swinging its massive head from side to side. Fortunately, as soon as it caught our scent it dropped onto all fours, turned and went crashing through the brush. I then realized that I had been so petrified by the

Introduction

encounter I hadn't even thought to pick up my camera.

"Not exactly a great start to our career as wildlife photographers!" I said to Sally.

We persevered, and our pursuit of wildlife led to many misadventures, from wading through boot-sucking swamps following moose, to the incomparable excitement of being mistaken for rivals by an angry elk. But we also shared many quiet moments with our animal neighbours. We petted porcupines (carefully!), entertained curious caribou, and watched a deer and her spotted fawn on a misty morning. Sally and I found that many animals sensed that we meant no harm and allowed us to approach surprisingly close.

With little more than a notebook, cameras and patience, we came to know birds and animals in a way few other people have experienced. Each time we saw something new and exciting we supplemented our personal observations by reading. In reading, we learned much about why animals behave the way they do. But we also discovered that many animals do not behave according to their textbook descriptions. Beavers and pine martens, for example, are classified as nocturnal, but after seeing them roaming at mid-day we realized that individual animals don't always follow the habits of their species. THEY certainly haven't read the books.

We also discovered that each animal has its own character and temperament: one moose was shy and ran from us with a long-legged trot, while the next was curious and edged closer to us as we watched, wide-eyed, from our canoe. And another moose charged our canoe without warning, spurring us to action. We backpaddled furiously, churning the water into a froth! Indeed, to record many of the stories and photographs we share in this book Sally and I risked assaults by weather, mosquitoes and even our subjects. All this, and more, is the story of our wild life as wildlife photographers.

TWO

Bugler of the Forest

One of the first bull elk that Sally and I tried to photograph lurched up from his bed in a timber thicket and charged me. I would never have imagined it possible to sprint with a cumbersome tripod and camera in one hand, and another heavy camera flopping on my chest. But I was so inspired that I did the one-hundred-metre dash, complete with two hurdles, in record time.

The stag thundered towards me, paused to thrash angrily at a bush that stood in his way, then let out a bugled call that made the hair on my neck stand on end. This was no ordinary elk. He was huge, heavy, and obviously agitated. His majestic branching antlers lay back over his shoulders almost to his rump as he bellowed, and his neck muscles were tense and swollen under a shaggy, chocolate-coloured cape.

This was also no ordinary time of year. It was mid-September and the two-month rutting season was well under way. In rutting season the normally shy and retiring bull elk becomes aggressive and irascible. A bull's goal during the rut is to round up a herd of females and defend his harem from all contenders.

Nose in the air, eyes red in the slanting sun, the stag that had chased me marched menacingly between the pines and out into the clearing. Sunlight glinted off his antlers and highlighted their polished, ivory points. The elk stopped again, raised his head to the sky until his antler tines touched his back.

"Eeeeeougghh!" His shrill call rolled through the wilderness valley.

A bugled reply echoed from somewhere deep in the forest. The bull immediately lost interest in us and remained motionless, his head turned towards the call.

After a moment, the bull stretched his muzzle forward and up, opened his mouth and bugled another ear-splitting shriek into the calm morning air.

"A-a-a-a-ai-e-eeeeee! E-ugh! E-ugh! E-ugh!"

The bugle sent shivers down my spine. It began as a wild, high-pitched shriek that might have been appropriate for a small creature, but seemed preposterous and undignified coming from the imposing elk. The shrill cry descended in pitch, then faded to a series of short, guttural grunts. With the effort of each grunt, the bull's belly and flanks pumped in and out.

Another reply echoed through the forest, antagonizing the bull even more. He took a few steps forward, looking intently into the forest.

From a safe distance Sally and I peered around a large tree. The challenger, a smaller animal, entered the clearing cautiously. This elk had a respectable six points on each antler, identifying him as a royal stag. The defender, whose massive antlers towered more than a metre above his head, had seven points on each side. He had clearly earned his designation as an imperial stag. In contests between two elk, the size of the antlers usually determines the dominant bull—and it seemed obvious to us that the defender had the advantage in this situation.

Bugler of the Forest

The larger bull slowly and deliberately moved across the clearing. With every strutting step the elk looked proud, arrogant, magnificent. He stood broadside to his foe and bugled, eyes bulging with rage. The intruder moved in, circled, then gouged the ground with his antlers, flinging clumps of turf into the air in a show of strength. The big bull bugled again, then charged—head down, great antlers forward—but stopped abruptly a few metres from the challenger.

The two stags circled each other for almost five minutes, posturing and tilting their heads to show their antlers to best advantage. Each bull mirrored the other's actions, pacing sideways, circling, and raking the ground with his antlers. Even as spectators, we felt the tension and drama of the confrontation.

Then, triggered by some action or sound we couldn't detect, they rushed each other and met with a clash of antlers. With heads down and antlers locked, the two pushed and twisted, their pounding hooves stirring up a cloud of dust. Suddenly, they broke apart. The big bull moved forward and shook his head, daring the intruder to continue.

The intruder took up the challenge and the bulls sparred, slamming together again in a shoving, grunting, antler-cracking contest of strength. This time the smaller bull's neck bowed and twisted until his brow tines touched the ground. He stumbled, but with a quick sidestep he was able to dodge the other's antlers. We expected the fight to end there, but even though the intruder was clearly at a disadvantage, he charged once more. The two animals locked antlers, straining until the larger bull drove the smaller backwards ten paces. The bulls broke apart again and, following some ancient animal code of ethics, the victor allowed the vanquished to retreat into the forest.

Fresh from his triumphant battle, the imperial stag made his way back to his harem. The cows displayed no interest—some had grazed in the meadow, while others had quietly chewed their cud through the entire performance. He approached one, sniffed her from behind and laid his neck on her rump. She moved away slowly. The bull followed, his tongue flicking in and out to reassure her that he was friendly, but it seemed she was just not interested in his advances. Speeding up, she broke into a trot and slipped into the forest.

During the following weeks we photographed two separate herds of elk, and each day the harem masters were challenged by other stags. Most encounters between bulls were brief and consisted primarily of bugling, posturing and displaying antlers. This allowed the bulls to size each other up and decide if they wanted to battle. If the bulls were evenly matched, a head-to-head shoving contest often followed. After less than a minute of back-and-forth pushing, one bull, usually the challenger, would break off the match and retreat.

Other stags were often attracted to the area by the sounds of bugling and clashing antlers. Although the larger stags only fought the defender one at a time, those waiting their turn added to the general mayhem by bugling and thrashing their antlers against bushes, driving the harem master to distraction.

The big bulls were also kept busy watching over their harem of cows and calves. It was difficult to keep the herd of ten to thirty animals together, even in clearings where the bull could keep watch over them. Young bachelors often tried to run off with a cow or two, and occasionally a cow just wandered away. Each time a cow tried to leave the harem the bull gave chase and cut off her escape route by blocking her with his body. If the cow was too slow to react she might get a nudge from the bull's great antlers. These herding tactics were quite effective in keeping the harem together, but the bulls had no time to rest.

With each passing day the bulls became more irritable;

being near them was exciting, to say the least. Even when not fighting, they seemed to keep themselves in a state of agitation by gouging the ground, shredding bushes and attacking small trees with their antlers.

A week after my first incident with a bull I was photographing another and felt more than uneasy when the bull stared intently in my direction. I turned my head slowly, checking on the location of the nearest tree, only to find that I was in the rather compromising position of being between the bull and one of his cows. He charged suddenly, and I retreated to the closest tree. There we stood for several long minutes, the bull on one side of the tree and me on the other. The elk was so close that I could see his bloodshot eyes and his nostrils flaring in and out with each steaming breath.

The bull began striking the tree with his antlers as I stood cowering on the other side, hoping the slender tree wouldn't break off and leave me facing the bull alone. His huge antlers curved around each side of the tree as he continued to vent his frustration on the branches. I expected to be skewered on the dagger points of a very angry stag at any moment.

Suddenly, a bugle from the other side of the clearing distracted the bull. It was Sally giving a most convincing elk call!

I took the opportunity to dash to the next line of trees. From the protection of the forest, I watched the bull give one last intimidating thrash and shatter the ten-centimetre-thick tree. I was relieved when the bull finally strutted back to his harem.

"Thanks," I mumbled, when I walked weak-kneed to where Sally was posted. "That wasn't one of my better encounters."

"Well, I hope you took some close-ups while you were there," Sally replied mischievously.

Still shaking, I confessed that I had felt no desire to take photographs of the raging bull's flaring nostrils.

Bugler of the Forest

With the constant challenges and little time to rest or feed, a dominant bull rarely remains strong enough to rule a harem throughout the entire rutting season. The challenging bulls often return to spar again, and sometimes win as the harem master becomes worn out from the constant vigil.

By mid-October of that year the intensity of the rut had diminished, and by the end of the month the season was over. The herd bulls were no longer in prime condition. They looked thin and tired; their heads drooped and their footsteps dragged. The haggard bulls hardly gave us a passing glance now, and appeared very different from the proud and arrogant masters that had controlled the large harems in September.

After the rut the older bulls gradually drifted away from the cows. The antagonisms of earlier in the season lessened and the bulls congregated in small bachelor groups to live together quite peaceably through the winter. The cows, calves and young elk formed their own herds of up to forty animals.

As snow began to fall in the mountains the elk moved to their winter range in the lower valleys. Elk will dig down to the grasses below if the snow is not deep, but mostly they will eat twigs of willow, aspen and poplar. We saw few elk through that winter, but as the snow piled deeper on the land we often thought of the difficulties they must have been experiencing. For many animals it is a constant struggle to find food and survive the long, cold winter.

But the elk seemed to winter well that year, and when we saw the herds of females in late spring many cows were accompanied by a young calf. The sun shone on the cows' reddish-brown coats as they grazed, and their tan rumps held our attention. The Shawnee Indians had named the elk "wapiti," meaning light rump, and as we watched the animals I thought of how appropriate the name was.

A persistent bull elk

Bugler of the Forest

One of the cows left the herd and slowly walked to a cluster of willows. Only then did we notice her calf lying there—its spotted coat provided the perfect camouflage for resting in the sun-dappled forest. The cow nudged the calf gently with her nose, and the young one immediately stood up on wobbly, broomstick legs to begin nursing. After only a minute or two, the cow terminated the nursing session by stepping over her calf and walking away. The calf hesitated, as if it wanted to follow its mother and continue feeding, but returned to the spot where it had previously bedded. The calf gave the area a sniff before dropping to its knees and lying down.

When we returned later that summer, some calves had grown so large they could no longer stand to nurse but had to kneel down to reach the udder. By the time the calves were two months old they were nibbling on grass and soon were grazing with the adults. But even in late August we saw calves trying to nurse. Occasionally a tolerant mother would allow one to feed briefly before walking away. If a calf was too persistent, the cow might strike it across the back with a front hoof or butt it on the side. This rebuff was usually sufficient to discourage the calf.

By September, the calves had outgrown their fuzzy spotted coats and were growing thicker, fuller winter fur. The adults' coats were also changing. Their tawny back and sides contrasted sharply with legs and manes which had turned dark brown as long guard hairs grew over the warm underfur.

Another autumn was approaching, and the stags that had ranged separately in small groups away from the cows and young were returning to the herds for another rutting season. The silence of the forest was again punctuated by the rattling of antlers on trees and the bugling calls of the elk.

Sally and I had now watched the wapiti through all the seasons and knew that autumn was the most dramatic.

Elk

We looked forward to another rutting season, a little wiser and somewhat more wary of the majestic bulls. We lay in our tent one frosty evening, listening to the calls echoing through the forest. Those challenging calls foretold of battles that would be fought. To us, the bugle of the elk is a stirring call of the wild. It rates with the howl of the wolf and the cry of the loon as one of nature's grandest sounds.

See colour photographs on pages 65, 66 and 79.

THREE
Porky

Petting porcupines can be prickly business—unless of course the porcupine is willing. I can think of at least 30,000 reasons not to walk up to a "porky" and pat it. Each one of those reasons is sharp, barbed and painful.

Sally and I had never known a porcupine closely until one April when a porky took up residence under the floorboards of a remote cabin we were staying in. Each evening we heard the slow, off-key humming of the porcupine under the floor, and occasionally we saw the prickly creature lumbering along the side of the cabin.

The first few times we approached, our new neighbour retreated at a clumsy waddle and hid under the cabin; but with each encounter the porcupine became more accustomed to us. After a week we could follow the animal as it wandered with its nose to the ground, searching for plants to eat. It was an amusing creature to watch as it ambled aimlessly from place to place with an awkward, pigeon-toed gait. Its fat body jiggled and the coat of quills rolled from side to side like a loose fitting saddle.

21

Porcupine

The porcupine's body was covered with thousands of quills which slanted backward and were partly hidden under a mop of long, blond guard hairs. The upper side of its tail looked like a pincushion filled with yellow-and-black needles. Even though the animal was heavily armed, our curiosity drew us a little closer each day. Sally even suggested it might let us stroke its quill-free nose.

"After you," I deferred. Although the porcupine had become quite friendly, I had no interest in finding out just HOW friendly it was.

Sally thought it over for a few days—it seemed like a risky thing to do, but the idea of petting our prickly neighbour appealed to her. She finally decided to give it a try. One morning, she crawled through the grass towards the feeding porcupine until she was less than a metre away. At the snap of a twig the porky bristled up, exposing each and every one of its quills. From Sally's location, it must have been a formidable sight and she froze in position, not sure whether to back away slowly or run!

She began speaking softly to the porcupine and gradually its quills returned to a relaxed position. After a few minutes Sally mustered up the courage to move closer. Approaching more carefully, she offered a freshly picked dandelion. This time the porcupine didn't bristle its quills or chatter its orange-coloured teeth in agitation, but gently took the flower. As the prickly animal nibbled on this delicacy, Sally gently rubbed its nose, then stroked its head (in the direction of the quills, of course).

The porcupine stared calmly with small black, unblinking eyes, its blunt nose twitching as it sniffed at Sally. Then it hummed softly with what we interpreted as pleasure. The porcupine moved closer, and rose up on its haunches to reach for another dandelion. With slow and gentle movements, it rested one paw on Sally's hand and gripped the flower with the other. Its paws were furred, but not quilled, and had very long, curved claws.

Porky

We were both thrilled at the porcupine's acceptance of Sally and amazed at its tranquillity. Sally later confided that she had felt anything but tranquil. During the encounter she had been tense and ready to leap back at any moment if the porcupine had quilled up again or swung its tail.

Sally and I often saw the slow-moving rodent around our cabin and had many opportunities to observe its daily habits. Like all porcupines, our friend's eyesight was poor. Even though its senses of smell and touch were well developed, it would often walk right by Sally or me without noticing us. Occasionally, the porcupine would stop in its tracks and rise up on its hind feet to get scent of us. The long guard hairs were so sensitive that when I touched the animal with the tip of a long twig, it immediately raised its quills and flipped its tail. Whap! The tail struck a log and left a number of quills embedded in the wood.

I hadn't expected such a lightning-fast reaction from the lethargic porcupine and I leapt backwards, prompting the animal to swat its tail against the log again. Sally was bent double with laughter at the sight of me scrambling in one direction to get away and the porky waddling as fast as it could in the other.

The quills ranged from five to twelve centimetres long, but only the tips had stuck in the log. I pulled one out and touched it to my finger—it was needle sharp, and I could imagine how effective and painful a weapon it could be. The fisher, a member of the weasel family, is one of the few animals that is successful in getting past the porcupine's armour without injury to itself.

The black area near the point seemed smooth in one direction and slightly rough in the other. Through a magnifying glass I could see each barb was like a fishhook, which would make the quills very difficult to remove. The rest of the ivory-coloured quill was smooth and ended in a short, weak root, which had detached itself

easily from the porcupine's skin. Porcupines do not "shoot" their quills, but because they are so loosely attached they will sometimes fall out or fly a short distance through the air when the animal whips its tail.

Our visitor seemed to be a loner, as most porcupines are, until early May when we heard new and unusual sounds coming from under our cabin. A few days later we discovered the cause of the noise—Sally and I were delighted to see a small, brown-haired baby accompanying the porcupine. No wonder she had looked so bulky! Porcupines give birth to one offspring between late April and June, and the young are born with eyes open and with soft quills which harden after about an hour. Young porcupines are quite mobile little creatures and can instinctively swat their quilled tails for protection within a few hours of birth.

Because they are such well-armed animals, we thought that motherly affection would be limited to nose rubs and grunts of endearment, but we often saw the young porcupine playing with its mother and climbing over her quite freely. The round little mound of silky-looking hair often followed its mother as she wandered from one patch of grass to another.

When the baby squealed for attention the mother responded by waddling to her youngster, then sitting on her haunches to let the baby nurse. The young one grunted and moaned as it fed, while the mother sat passively, eyes closed and one paw resting on her youngster's head.

By the time the young porcupine was three weeks old, it nursed less frequently and we often saw it foraging alongside its mother. They both searched for their meals by smell more than sight and seemed to prefer dining on the early summer growth of dandelion and clover to eating grass. We also saw them feed on the buds and leaves of willow bushes and young conifer trees. The tender new growth must have been a pleasant change

Porcupine

from the inner bark of spruce and pine trees, which is a porcupine's usual diet most of the year.

Although we were pleased to get to know the porcupine family, we discovered they had several bad habits. The most irritating was their insatiable craving for salt. Sweat-soaked axe handles, boots and canoe paddles were favourite snacks, so we had to remember to keep these out of reach. One morning I found the mother porcupine contentedly chewing on a work glove that I had left by the woodpile. Keeping in mind that it is not a good idea to anger a porcupine, I approached slowly and ever so gently tugged the glove from her grasp, offering a tender dandelion in return.

Sally and I also found that porcupines are not silent creatures. We were often wakened in the night by grunting, moaning and humming from under the cabin. The mother had an amusing habit of humming to herself as she wandered from place to place. After listening to their noises I practised my own version, and during the summer I became an accomplished porcupine caller. I would hum softly, working slowly down the scale for about four notes, and the young porcupine would come waddling toward me. The first time its myopic eyes discovered that I was not another porcupine, it stopped in its tracks, snorted with surprise, then quickly lumbered away. Later in the summer it became as trusting as its mother and would come when I called, to see if I had any treats of freshly picked fireweed.

Not all porcupines are as approachable as the mother and young at our cabin. Most of our encounters in other years had resulted in the porcupines bristling their quills, chattering their teeth and turning away from us. If we approached closer the porcupine usually crouched with head bowed, legs stiff and back hunched, confident in its protection and ready to lash out with a spiked tail.

One porcupine we came across surprised us by climbing a tree with a skill and strength that seemed foreign to

"Porky"—an accomplished tree climber

its sluggish nature. It stood on its rear legs at the base of the tree and reached up with its right paw, then left paw, and sunk its claws into the bark to grip the tree firmly. The porcupine next swung both hind legs upward beneath its body, and from this new hold reached up again to repeat the action. In this way, the bulky creature moved quickly up the tree, then gave us a backward glance and settled in a fork of a branch for a rest.

An hour later we returned to find the porcupine inching its way down, holding its bristled tail against the tree for support. In contrast to its confident climb, its descent was slow and awkward.

We were also surprised to find that, in addition to being good tree climbers, porcupines swim quite well. During one canoe trip we saw a porky walking along the shore of a lake. When it came to a fallen tree which obstructed its path, the porcupine waded into the lake and swam around the windfall. Its hollow quills must have helped it remain buoyant, because it floated quite high out of the water. The porcupine was a strong swimmer and seemed to be at home in the lake, although it took great care to hold its tail well out of the water.

Porcupines normally don't wander far from their home range. However, during autumn they become restless and roam further afield as they begin looking for mates. This is the time of year when porcupines are most vocal, and we occasionally heard the crooning and whining sounds from females that were answered by the males in louder tones.

When porcupines meet at this time of year they check one another's scent, and often touch and rub noses. If two males meet they will challenge each other vocally with grunts and groans, but these encounters rarely result in fights. The two just continue wandering on their way, looking for a receptive female.

When a male and female finally find each other, they may remain together for some days before mating. We

Porky

were fortunate to see part of their courtship ritual one afternoon. When the two porcupines met, they stood on hind legs and placed their paws on each other's shoulders. The porcupines looked as though they were dancing as they rocked from side to side with a slow rhythm, lifting one foot and then the other for several minutes before wandering down the trail together.

Because the porcupine is such a prickly creature, its mating habits have long been a subject of wild speculation around campfires. One northern trapper we met shared several imaginative theories of how porcupines mate without stabbing each other. The truth is, the female pulls her quills flat against her body and arches her tail over her back to provide a soft, quill-free cushion on which the male can lean. Needless to say, the female must be more than agreeable to the male's approach—even a porcupine must be careful when it is with another porcupine.

After mating season the porcupines go their solitary way again. We have seen few of these animals during the winter months because they spend most of the winter eating bark high in trees, where their dark hunched-up form is difficult to spot. Porcupines do not hibernate and may remain in one tree for weeks at a time, only coming down to a den if the weather turns extremely cold.

I recall seeing a porcupine waddling through the snow one winter, leaving a snaking, s-shaped trail as its tail swung from side to side. The temperature was well below zero but the porcupine seemed quite comfortable with its thick layer of underfur protecting it from the winter's chill. It moved slowly, as usual. With 30,000 quills for protection, a porky is one animal that doesn't have to hustle in any season!

See colour photograph on page 71.

FOUR

Black Bear Ahead!

There are few phrases that can raise my heart rate and shorten my breath as quickly as an urgently whispered, "Bear ahead!"

On one such occasion, I was following Sally over the crest of a hill when she suddenly stopped in her tracks and uttered those words. Not twenty paces ahead of us, standing as still as Sally was—and carrying the same startled look on its face—was a bulky black bear.

Any large wild animal can be dangerous if surprised or frightened, and this bear, frozen in mid-stride, looked as surprised and frightened as we were. Nervously, I considered our options: Should we back away slowly? Stand still and hope for the best? Or shout and clap our hands to scare the bear away?

Sally and I remained motionless, while the bear seemed to grow even larger as the hackles rose on its neck and shoulders. I cringed when the bear opened its mouth, exposing a hint of huge white fangs.

"Black bears are ninety percent vegetarian," I

Black Bear Ahead!

whispered, trying to reassure myself. The bear rose up on its hind legs for a better look at us with its nearsighted eyes; then its long nose twitched as the bruin tested the wind for our scent. The animal stood almost two metres tall, and its huge front paws with black claws hung by its sides. The bear's thick coat glistened in the sunlight, adding to the intimidating image.

After several moments of mutual staring, the bear let out a "woof" that almost lifted me out of my boots, then turned, dropped to all fours and loped away. Sally and I let out a simultaneous sigh of relief as the large, jiggling rump disappeared from view into the willows. Judging by the speed it was moving, the bear didn't plan to stop until it reached the other side of the valley.

"Should make a great campfire story," I said to Sally when our heartbeats returned to normal.

"Right... if we halve the distance, double the bear's size, and embellish the story with a few ferocious growls," Sally added. I had to admit, a meeting with a bear that scrambled the other way without charging lacked the dramatic punch-line necessary for a fireside story.

Although not many of our bear encounters have been of the fireside-thriller variety, they certainly were exciting. Bears are usually shy creatures, but they can act in unpredictable ways. We knew that if we approached too closely, a bear would either hurry away or charge. Neither reaction would be to our advantage as photographers.

One morning, we followed a big male cautiously for more than three hours as he wandered in search of food. Our desire to take bear photographs was tempered somewhat by a healthy fear. We moved slowly and kept in full view of the animal, carefully watching his reaction to our presence. The bear threw an occasional glance in our direction, but his ears and hackles remained relaxed—a good sign. I became worried at one point, however, when

Black Bear

the bear stood up on his hind legs, slowly swinging his head from side to side to sniff the air currents.

"Let's give him more room," I whispered, taking a step back. The bear must have been satisfied that we were neither food nor foe, because he returned to all fours and proceeded to meander across the slope. This bear seemed willing to tolerate our presence as long as we continued to use good judgement and a long telephoto lens.

The bruin shuffled along, nose to the ground searching for food. He travelled at an unhurried two- or three-kilometre an hour pace. Bears have a clumsy-looking, shuffling gait as a result of placing their feet flat on the ground, rather than rolling from heel to toe as people do.

The slow-moving bear was involved in what appeared to be brunch on the move, nibbling at a clump of grass here, a flowering plant there, then rambling on to investigate a rotting log. The animal tore apart the log by hooking his claws into the wood and pulling it towards his chest. This effort was rewarded with only a meagre meal of ants and other bugs, which he quickly licked up. It was comical to see the huge beast working with such fervour in order to dine on such few insects. He took another great swipe at the log, sending splinters of wood flying through the air in a cloud of dust, then thrust his muzzle into the log and scooped up more insects with his long, pink tongue.

Black bears, while not as large as grizzlies, are still big animals. A mature male can attain a weight of more than two hundred kilograms, and when standing upright they can tower more than two metres high. They are also strong, and we had a chance to observe this bear's tremendous power when he pushed aside a boulder which Sally and I, working together, would not have been able to budge. Using his powerful front limbs, he tugged and pulled, first with one paw, then with two, and finally succeeded in rolling the large boulder over. A cursory inspection of the damp hollow with his nose seemed to

reveal nothing edible, so the bear continued on his way.

Then something attracted the bear's attention. He perked up his ears and twitched his nose, and made a leisurely side trip to satisfy the insatiable curiosity which is characteristic of all black bears. The detour brought him to a colony of ground squirrels. With lightning-fast speed he lunged for the nearest squirrel. We were amazed at the quick reaction of the bear, who had at first appeared to be a lumbering, clumsy creature. But he wasn't quite fast enough to catch the ground squirrel, who dove into the safety of its burrow. Although the bulk of a bear's diet consists of vegetation, they will eat almost any small animal they can catch, as well as fish and carrion.

A short time later we watched the bear pause before a cluster of flowers. We were amazed that an animal with the strength to move huge boulders also had the patience and skill to nibble delicately on flowers. We saw that the bear's lips could be manipulated with amazing dexterity and were especially useful when feeding on the blossoms. Carefully tilting his head, the bear singled out a cluster of flowers with his lips, then nipped the blossoms neatly off the stalk.

By late morning the temperature had become uncomfortably warm and the bear wandered into the forest, searching for a cool place to snooze. Because of their thick, dark coats, bears do not like the hot sun and are most active during early morning and late evening. Sally and I agreed it would be too risky to follow this bear into the heavy timber where we would have difficulty keeping an eye on him.

Most bears, like the one we had followed, are solitary creatures. Adults seek the company of others only during the short mating season of early summer, although females over three years old are usually accompanied by cubs. Cubs are born in late January or February and stay with their mother for sixteen or seventeen months. During this time the cubs are taught all the skills they need to

Black Bear Ahead!

survive, from foraging for food to climbing trees.

Sally and I were able to watch the interaction between sow and cubs one autumn afternoon when I detected a movement in some bushes a short distance from the trail. Looking more closely, I spotted a small black cub scrambling onto a boulder. I glanced apprehensively over my shoulder, for where there is a cub there is sure to be a sow, and the worst possible place to be is between the two.

Suddenly, a large bear leapt to the top of the boulder. With a grunt and one sweeping cuff she sent her cub flying through the air to land in a tangle of brush below. I couldn't help wondering why the cub received such rough treatment; perhaps the sow was just batting her youngster to safety. The sow turned and gave us what I interpreted as a warning look, then plunged into the thicket herself. For several moments all we could see was a stormy tossing of limbs and willow tops as the bears ploughed away from us through the tangle.

The sow finally emerged some distance away with two cubs in tow, one pitch black like herself, the other light brown. Black bears, despite their name, may be black, brown, cinnamon or even blond in colour. The young bears were playful creatures who tripped and tumbled head over tail and tackled each other until they were brought into line by their stern mother.

She led the twins up the hillside at a fast lope. It was then that I decided it would be foolish to try to outrun a bear—the trio of bears travelled two hundred metres straight up the slope at a speed faster than we could ever have run and stopped only when they had reached the sanctuary of some tall trees. The cubs promptly scurried up one of these while their mother sat guard at its base. Black bears are accomplished climbers, equipped with five short, curved claws on each foot. The youngsters climbed with a series of quick bounds, grasping the tree with their front claws and pushing with their hind legs.

A burly black bear

Black Bear Ahead!

The cubs remained up the tree for almost half an hour, resting on a branch with their feet dangling lazily down on each side. At a signal from the sow, the youngsters came inching backwards down the tree and dropped the last two metres to the ground. Following their mother, they quickly disappeared into the underbrush, apparently unshaken by the fall.

Bears are creatures of habit and travel regularly on familiar and often well-defined trails in their home range of fifteen to twenty square kilometres. We encountered the sow and cubs a few days later in a blueberry patch near where we had first seen them. The lush berry patch must have seemed like paradise to the bears. This time the sow was more interested in dining than in charging us or running away. The bears were up to their muzzles in blueberries, feeding voraciously in preparation for winter.

Bears take the easy way of storing food—on their bodies in the form of fat. With such a good crop of berries to dine on, the trio we had encountered would be ready to retire to their dens before the first snowfall blanketed the forest. Bears are not true hibernators, but spend the winter in a deep sleep with their body temperature and heart rate just below normal. Even though bears snooze through the winter, they occasionally wake up during warm spells and wander about, then return to their dens.

Our encounters with black bears during their active months had certainly added plenty of excitement and suspense to our wilderness outings. But from each October until the following March or April there was little chance of meeting one of these animals face-to-snout. With the bears in their dens during the winter, the forests certainly seemed a little quieter—and tamer.

See colour photograph on page 66.

FIVE

Woodland Drummer

The most enchanting woodland sound Sally and I have ever heard was the hollow drumming of two ruffed grouse one evening during a northern spring. The weather was clear and cool, prompting us to spread our sleeping bags on the forest floor instead of setting up the tent. As a full moon crested the mountains, the first hollow booming of wings echoed through the forest. After only a minute or two a challenge sounded from another direction.

Each drum roll rose to a crescendo, reverberated through the forest, then diminished to a low thrum. To our delight the two grouse drummed back and forth for more than an hour. We listened, spellbound. The drumming seemed to come from everywhere, yet nowhere, and had an almost supernatural quality. It was an appropriate sound for the ethereal mood of that silver-white moonlit night.

Sally and I were wakened early the next morning by a repetition of the evening concert. We scrambled out of

Woodland Drummer

our sleeping bags hoping to find one of the drumming birds so we could watch its performance. The hollow sound was almost impossible to pinpoint. Although it had seemed very near, we walked for more than half a kilometre before the drumming became louder. Then, as we neared the source, the forest became silent.

Sally and I combed the woods each morning for several more days, but each time we approached a drumming grouse the performance ceased. We found a drumming area quite by accident the fourth day when a grouse exploded from the brush with a whirr of wings. When our heartbeats returned to normal, we saw that right in front of us, partly hidden by willows, was a moss-covered log with a worn area at one end.

"This must be a grouse drumming stage," I whispered to Sally. It was an exciting discovery, and we made plans to return and watch the drumming display.

At dawn the next morning I sat quietly and peered through a tangle of willow branches. After waiting for almost an hour, I finally heard the soft rustle of dry leaves. Moments later I saw the bobbing form of a mottled-grey bird stepping carefully along the ground. Holding my breath in anticipation, I watched the ruffed grouse move forward cautiously, pause to look for signs of danger, and then jump to his stage.

The cock grouse strutted back and forth along the log. He was a striking bird with his crest of head feathers fully erected and an enormous black ruff encircling his neck. The grouse was about the size of a small chicken; his feathers were splashed with a mixture of black, grey, brown and white to blend with his surroundings.

A ringside seat at a grouse performance is a rare woodland priviledge. I sat entranced as the grouse paraded with his wings held low and flight feathers spread wide. Then he stood still, head and neck stretched to display his ruff of feathers, which he fluffed out to full splendour with a shake of his head. His barred tail was raised and

A male ruffed grouse on his drumming log

Woodland Drummer

fanned to an almost perfect half-circle, showing off a broad band of black at the end. The bird then dug his claws into the log and pumped his strong wings forward twice, seeming to test them.

"Boom.... Boom."

With tail fanned and braced against the log for balance, he pumped again, his wings moving faster and faster, picking up a vibrating rhythm as he increased the tempo.

"Boom... boom... boom-boom-boom-brrrrrrrmmm." The hollow drumming echoed through the forest. The sound was surprisingly soft at close range, even though we had been able to hear it from more than half a kilometre away.

At the conclusion of the drumming, the grouse gradually relaxed and smoothed his feathers down, then peered intently all around. I was not sure whether the grouse's piercing black eyes were looking for a rival, for a receptive hen, or at me!

By my watch, he beat his wings every five minutes, paying little attention to my presence. But between drumming sessions the grouse was very wary, and when I squirmed to adjust my uncomfortable position he dropped immediately from the log. With a highly indignant "tut, tut, tut, tut," the grouse walked away and silently disappeared into the woods.

Hoping the grouse would return, I decided to wait. I took the opportunity to study the drumming log—it was large, old and almost level. I was sitting on the ground and could see that from the grouse's vantage point there was an unobstructed view for several metres into the forest. Overhanging branches offered protection from winged predators and there was plenty of thick brush nearby where the grouse could hide. On one side of the log there was a neat pile of pellets indicating that the grouse had used this log for many performances and that he always faced the same way.

I carefully set the camera on the tripod, attached the

Ruffed Grouse

appropriate lens, then focused on the drumming log. After a few minutes of silence, just as the light was becoming bright enough for photography, the grouse resumed drumming—from a new location. It was incredibly frustrating to have finally found a drumming log, but to have missed the grouse ... a rather important element.

I knew from experience that it would be impossible to try to sneak up on the bird, but I had read that they ardently defend their favourite drumming log and territory against any other male grouse. Having nothing to lose, I thumped my chest with an open hand, hoping to imitate the accelerating rhythm of the grouse. All was quiet in the forest. I beat my chest again and waited to see what might happen. My ruse was successful; within minutes the grouse came strutting stiffly to challenge the rival he must have thought was using his special log.

The grouse stopped, surveyed the area, then, without the usual preliminaries, hopped onto the log and began drumming. He quickly placed his feet in position, pressed his tail against the log and extended both wings backward. A few single beats were given and then the wings began to flash—faster, faster, faster—until they became a blur. Finally, they slowed and, after about six seconds, stopped.

This time I was able to take many photographs and closely observe the drumming. Indians called the ruffed grouse "the carpenter bird" because they thought it drummed by beating its wings against a log. The sound is actually created by the compression of air as the wings are brought forward rapidly.

Male grouse drum to attract females to their territory. Although grouse might perform at any time of the year or at any hour of the day, the regular drumming season starts in March and declines in May. For all his effort, the grouse's performance seemed to be in vain that day. No male grouse responded, nor did any females visit.

I was surprised at his lack of success because Sally and

Ruffed Grouse

I had seen several females in the area; they can be distinguished from males by their smaller size and the incomplete black band on their tails. Each time we saw the hens they appeared to be ignoring the drumming around them. They were busy "budding"—pecking at the new buds while they balanced precariously on the slender outer branches of willow bushes. Grouse will eat almost any greenery as well as nuts and berries, but the willow buds seemed to be a favourite at this time of year.

The enthusiastic drumming of the males must have been productive after all, because in mid-June we came across a hen with her brood. She was under the low branches of a spruce tree, softly cluck-clucking to her chicks as we approached. The hen uttered a short "breeet," and three balls of downy fluff immediately scattered for cover. Their mottled brown colouring provided such excellent camouflage that I was able to locate only one of the three.

The hen left her chicks and approached Sally and me with anguished clucks as she limped, her left wing trailing "uselessly" behind. It was a most convincing act, meant to draw our attention away from her brood. She then fluttered in the opposite direction of the chicks and went through her injured-wing routine again. She seemed a ready meal for any predator, but when I approached she lured me even further away then burst into the air on two well-functioning wings.

Through the summer we watched as the chicks grew into slightly smaller versions of their mother. By September the young began to disperse through the forest as they searched for their own homes.

One afternoon Sally and I came across a young male grouse resting under a willow bush on his journey to a new territory. "Good luck," I whispered under my breath. I hoped that he would survive the rigours of winter to enliven the woods with his own drumming in the years to come.

See colour photograph on page 68.

SIX

Curious Caribou

Across the frozen lake several woodland caribou lay in the snow, their antlers set in dramatic relief against the white backdrop. Each animal faced a different direction to improve the herd's chances of detecting any approaching predator, especially the wolf, their most feared enemy. This made it especially challenging for us to move to within camera range without spooking the herd.

"How about trying to impersonate a caribou... their eyesight is so poor, we might just get away with it," Sally suggested.

It seemed like a crazy idea, but I could think of no better way to get close to the herd. Feeling slightly foolish, we each bent forward at the waist, our backs parallel to the ground, and skied towards them. Sally led, her arms held high like a pair of antlers, while I crouched behind, doing my best imitation of a caribou's back end.

As we moved closer the animals stood up and shuffled uneasily. They kept glancing at the trees, as if they wanted to run for cover, but seemed held in place by their

Caribou

curiosity. It was clear that they didn't know what to make of us: their ears strained towards us and their eyes bulged in utter disbelief. Several of the young were actually overcome by their curiosity and broke from the herd, trotting towards us eagerly. Then something changed their minds—either a signal from their parents that we couldn't detect or a clearer view of us—and they dashed back to the safety of the group.

There were thirteen animals in the herd, an average-sized group of woodland caribou. All caribou are gregarious and seldom travel alone, but unlike their tundra cousins, who travel in groups numbering thousands of animals, these caribou rarely travel in groups of more than twenty.

These caribou were clothed in their thick winter coats of hollow hair; even their muzzles and short tails were covered with hair for warmth. Both males and females had mahogany-coloured antlers, although the antlers of the males were more massive, with a greater number of tines. The bulls' antlers swept back and up in an arc over their heads and featured a shovel-like brow tine over their muzzles. Even though the bulls stood not much more than a metre high at the shoulders, the combination of large antlers, thick coats and flowing white manes gave the animals a striking and majestic appearance.

At a distance of twenty metres I set up my tripod and focused the camera. This procedure seemed to intrigue the caribou even more. Caribou will instinctively run from animals they recognize as predators, but these caribou had probably not seen humans before and were curious about this new object in their lives.

"What a picture!" I couldn't help exclaiming, excited by the image in the viewfinder. At the sound of my voice, the caribou turned in unison and trotted away, long noses held level with their backs and antlers curving gracefully over their shoulders—a haughty, regal pose.

Just as Sally and I were beginning to despair of *ever*

Caribou

photographing a caribou, they stopped, turned, and came back for a second look. This time I kept my comments to myself.

We stared. They stared. We moved back. They moved forward. I whistled a ragtime tune and they came even closer, then hoisted their stubby tails, gave a lighthearted skip and trotted away, only to return again for another look.

This was one of those rare occasions when the subject of a wildlife photograph came back to give us a second chance (not to mention a third and a fourth). All too often our subjects move before we have a chance to focus; even if they pose obligingly, the position of the sun often makes photography difficult. But everything was right that afternoon: good light, cooperative subjects and plenty of time to compose each photograph. I even had time to change lenses, fiddle with f-stops, and change a roll of film while they stood as if posed for a group portrait. We owed our good fortune entirely to the caribou's famed curiosity.

After five minutes of staring, one of the bulls either tired of us or decided that we were, after all, something to fear. Abruptly, he pivoted and raced along the lake. The rest of the herd followed his lead. This time they turned into the forest without looking back.

Sally and I were left on the expanse of trampled snow with our cameras and tripods, feeling strangely abandoned. We'll never know what triggered their departure.

The next day we followed the caribou trail for several kilometres through the forest, hoping to learn more about their habits. Although the animals kept more or less together, each caribou found its own way, pausing now and again to dig through the snow and expose the vegetation underneath. Because winter is a lean season, caribou are constantly on the move in their search for food. They eat whatever forage is available, from lichens, grasses and mosses to twigs of willow, dwarf birch and other shrubs.

Curious Caribou

We saw several pits where the animals had scraped through deep snow; they must have expended an incredible amount of energy to obtain a few mouthfuls of food. It used to be thought that caribou used the "shovel" on the front of their antlers to dig through the snow, but the brow tine rarely extends past the nose and would be of little use for digging. Actually, they use their broad hooves. "Caribou" comes from an Algonquin Indian word meaning "shoveler of snow." An apt name, I concluded, looking at the many deep craters.

In our travels we often found the meandering trails of caribou, but our next encounter was not until early June. We discovered a small band of cows and young calves hidden among the trees part way up a mountain slope. The cows seemed more nervous than during the winter and paced back and forth, ears cocked and nostrils flared, searching for scent or sound. At this time of year the cows are worried about their newborn calves, which are easy prey for wolverine, wolf, bear and lynx.

The cow nearest us must have caught our scent. She lifted her head high, ears pointing forward. Then she raised her tail and stretched one hind leg out to the side. This rather odd-looking pose warned the other caribou of danger. She began moving away, but her calf tottered towards us on its spindly legs. The fuzzy-coated animal had taken only three steps when its mother grunted a low-pitched "a-a-a-w." The calf paused to glance at its mother, but did not rejoin her until the other adults decided to bolt.

Like all caribou calves, this one was not spotted like other young of the deer family. Its coat was a dull dun colour, except for its muzzle and legs, which were black. And like all caribou, young and full-grown, it had feet that seemed ludicrously large for its body. Despite their oversized feet, caribou run with a smooth, rocking gait and can attain speeds of sixty kilometres an hour for short distances. For sheer poetry of motion, there are few

Woodland caribou returning for a second look

Curious Caribou

animals more graceful on the run.

Even this two- or three-week-old calf was agile and fleet-footed. Caribou calves are up and running within hours of birth and can outrun most predators when they are only a few days old. We watched as the calf negotiated a steep snow slope then continued up the rugged mountainside, a climb so steep we would have expected to see only mountain sheep or goats tackle it.

Woodland caribou usually travel only ten or twenty kilometres from their winter range in the forest to a summer range in the alpine areas, but they are restless creatures and are constantly on the move. That summer we saw small bands of cows and calves wandering across the high slopes, where they grazed on a wide variety of sedges, grasses, flowers and brush. The older bulls were nowhere to be seen; they preferred to wander alone or in small groups lower on the mountains until autumn, when they would rejoin the females for the rut.

Only once did we see the restless caribou remain in one location for any length of time. In August we came across a herd gathered on a high, windswept ridge. They stood quietly on the snowfield, heads low, looking exhausted. Suddenly, one caribou rushed off at full speed, as if chased by demons.

These "demons" were mainly botflies trying to lay larvae in the nasal passages, or warbleflies laying eggs on the thin summer coats of the caribou. The rest of the beleaguered animals stamped their feet, twitched their hides and held their nostrils to the cool snow for relief. When the caribou could not stand the torment any longer they would run madly across the snow in a vain attempt to elude the flies. After thirty metres of crazed running, the miserable animals would stop and bury their noses in the snow again. The caribou were so preoccupied with the persistent flies that they completely ignored our approach.

Suddenly one of the fly-crazed animals raced to a

nearby snowfield. The rest of the herd followed but stopped part way to have another look at Sally and me. This was a good opportunity to try a trick an old woodsman had shared with us about how to attract caribou. Before the caribou could turn away, Sally lay on her back and began pedalling her legs in the air.

It worked! The ever-curious caribou came closer, watching Sally intently. They took a few steps, looked our way again, and began eating grass. Each caribou took a nip here, a nip there, as they wandered closer, casting glances our way every few bites. I had to suppress a chuckle—it almost looked as though *they* were trying to sneak up on *us*. The caribou were now within fifteen metres as I peered through the viewfinder of the camera.

"Keep going," I encouraged Sally, when I noticed the pace of her pedalling had slowed.

"Okay... okay... the things I have... to do... for photographs," she said between breaths.

Sally and I smiled at each other. Despite her answer, I knew that she was thrilled to have the caribou approach us so closely. In fact, there were probably few places she would rather be than with caribou on a sunny mountain top.

See colour photograph on page 67.

SEVEN

Little Haymaker

After a long, hard climb to the high country, Sally and I stretched out in a meadow for a short rest. I lay with my eyes closed until a faint rustling caught my attention. Slowly I opened one eye and met the stare of a tiny pika... only an arm's length away.

Except for its long whiskers, which quivered slightly, the pika sat perfectly still, watching me with an unblinking gaze. It was an attractive little creature, with soft, grey fur and large round ears, and was so small it could have crouched comfortably in my hand. After only a few seconds the pika scampered to the shadow of a boulder, where it gave a shrill, high-pitched "eeenk," then disappeared from view. Knowing there were endless passages under the rockpile, I didn't even try to guess where the pika had gone.

The pika must have been used to its predators giving up easily, because it reappeared in minutes and began moving about the rockpile. As long as we stayed perfectly still, the pika was content to ignore us and to go about its

53

business. The small creature scooted back and forth over the rocks like a wind-up toy: its legs were so short we could barely see them propelling the animal. Although the pika is a member of the rabbit family, pikas don't hop like rabbits and they look more like mice without tails.

Sometimes the pika would disappear from sight for several minutes. We soon discovered the animal was making frequent trips to an alpine meadow which bordered the rockpile. We watched the pika roam from plant to plant, clipping greenery with its teeth. With a mouthful of this newly cut "hay" carried crosswise in its mouth, the pika raced to its haystack under a large boulder.

The pika's method of gathering plants was amusing to watch. When the plant was small, the pika gripped the stem between its teeth and pulled backwards with all its might, until the stem or root gave way, sending the small creature tumbling backwards. When the plant was larger, it cut the stem with its teeth, then dragged it over the rocks and boulders, all the way to its haystack.

The haystack was in the shelter of an overhanging rock, but in a location where the afternoon sun could reach it. The pika kept several haystacks, adding fresh material as soon as the previous layer was sun-dried. We checked out one hoard and found an amazing variety of plants. In addition to grasses, there were small twigs, willow leaves, fireweed, aster and heather.

All summer long the pika literally makes hay while the sun shines. Unlike marmots, who feast during the summer to fatten up for their hibernation, pikas are active all year and must store sufficient food to last through the winter months. The pikas' small size belies their rugged nature; their round shape, short legs and hidden tail expose little to the cold, and their thick fur provides good insulation.

We heard several other calls and found that the area supported many pikas. They were superbly camouflaged as they rested on top of sloping rocks. Only when a small

A charming pika

Pika

grey rock moved did I realize that my eyes had passed over a pika without recognizing it. Other pikas scurried back and forth along well-worn paths between the meadow and their haystacks.

Sally and I were so intrigued by the little haymakers that we set up camp nearby, hoping to photograph them. They seemed to be constantly on the run between the meadow and their haystacks; and after a while we noticed each pika followed a particular route. The pikas inspected their haystacks several times a day and often roamed around what seemed to be the perimeter of their territories. They also found time to take short siestas in the mountain sunshine.

We spent the afternoon studying the pikas' habits to find the best time and place to take photographs. The next day Sally chose to wait for her photographs until an animal near her took a siesta, while I posted myself near a frequently visited haystack. When I stepped closer, the pika froze in position and began calling frantically. Each time the pika called, it thrust its head forward, opened its mouth wide and twitched its ears, all at the same time. As the nasal "eeeenk" was repeated, each succeeding one became progressively fainter, giving the impression that the pika was moving away.

The pika was so sure of its camouflage that it continued to squeak as I set up the camera and tripod. I focused, then gently squeezed the cable release. At the sound of the shutter the pika leaped into the air with a squeak, swapped ends, and scurried across the rockpile.

While I waited for the pika to return, I heard frequent calls from other pikas on the slope. It seemed that they called to defend their territories and haystacks from other pikas, usually sounding their one-note alarm at such volume that it startled me, even though it was not unexpected.

Pikas call not only to defend their small territories but also to warn each other of predators. When the shadow of

Little Haymaker

a hawk passed over the rocks, we heard a series of loud cries from the colony. Each pika called a warning to its neighbour. But later, when a weasel ventured near, the pikas fell silent. Because weasels can easily enter the rocky crevices where pikas live, the pikas took care not to betray their location, even by sounding an alarm.

That evening brought a heavy rainstorm and high winds. We were awakened several times through the night by the drumming of rain on our tent and the incessant calling of pikas.

"This will teach us to pitch our tent right beside a pika colony," Sally mumbled at three in the morning, when a particularly vociferous pika was preventing us from returning to sleep.

When daylight came we found that the pikas had moved each haystack to a safe, dry place even further under the rocks. The pikas must have worked all night to protect their winter food from the rain.

When the sun finally broke through the clouds the pikas began bringing their hay into the sunshine to continue curing. Pikas are incredibly industrious creatures: by the end of a summer, each tiny pika would have the required stockpile of about ten kilograms of dried hay for winter nibbling.

We admired the resourcefulness of the pikas. With an extensive network of tunnels and an early-warning system, they had little to fear from enemies. The pikas stored enough hay for a comfortable winter, and on quiet, sunny days they even took time to sunbathe on a favourite boulder. These little haymakers certainly knew how to make the most of the high, wild country of the north.

See colour photograph on page 71.

EIGHT

Winged Hunters

Far across the marsh, a low-flying bird cruised just above the waving crests of grass. The winged hunter crossed and re-crossed the open area as it investigated every contour below for sound or motion, ready to pause and pounce.

Sally and I stopped paddling the canoe as the bird flew closer. As it passed in front of us the long-winged bird banked to one side, revealing a white patch above its tail.

"A northern harrier," Sally said, peering through the binoculars. The conspicuous white rump patch was the key to identifying the harrier. A moment later, Sally reported that it was a female: the bird's feathers were dark brown above and its pale underside was streaked with rusty brown. Males are grey above, mostly white below, and sport black wingtips.

We were entranced by the harrier's graceful, buoyant flight as it languidly performed a variety of slow pirouettes and smooth sideslips on the wind. The graceful bird occasionally renewed its speed with a few deliberate

Winged Hunters

wingstrokes, then glided for almost one hundred metres. With wings held in a shallow V, it tilted its body from side to side, surveying the countryside.

Suddenly, the harrier stopped in mid-air and plummeted earthward. A moment later, the bird ascended with a tangle of brown grass in her talons. But the grass was dropped after a few metres and the bird continued its tireless hunt for a meal. We guessed that the ball of grass had been a mouse nest; perhaps the harrier had been hoping for a meal of young mice.

Further along, the bird paused again, skillfully holding its position by sideslipping and stalling in order to harry its prey and counter every dart and turn of the prospective meal. The harrier must have sensed a moment of indecision, for it suddenly dropped like a stone, with wings raised and talons outstretched to snatch the prey.

The harrier missed its target and rose at once to hover a metre above the grass. After only a moment, it dropped again in a twisting dive. This lunge must have been successful, for the bird remained in the grasses, hidden from our view for some time. Usually, harriers devour their meals on the ground, although occasionally we had seen a harrier carrying a mouse in its talons as it flew to some convenient stump or to its young at a nest.

The persistent, low-level attacks of this hunter are the reason that it has been named a harrier. The kinds of prey most vulnerable to this method of hunting are mice, voles and other small grassland rodents. Snakes and young birds are also taken occasionally.

The northern harrier often circled near us as it crossed the marsh. During one pass, I pursed my lips against the back of my hand and let out a squeak that sounded like a mouse. My second squeak produced an immediate reaction—the hawk dropped one wing and pivoted, soaring towards us. It flew silently overhead, tilting slightly from side to side in the wind.

Sally and I could now see its features in greater detail.

Winged Hunters

The bird had a ruff of feathers on the sides of its face, giving it an owl-like appearance. As with owls, this funnel of feathers directs sound toward the ears and helps account for the bird's exceptional hearing. Guided by sound alone, a northern harrier can detect a mouse at a distance of five or six metres.

This harrier would probably cover more than one hundred kilometres a day with its graceful flight, even though its travels would be confined within the two or three square kilometres of its hunting range. The northern harrier used to be called a "marsh hawk," but this name was inappropriate because its haunts are not limited to marshes. We had often seen harriers soaring over meadows, grasslands and even high alpine country.

Once the bird had moved beyond the range of our binoculars, Sally and I left our post and canoed along the adjacent lakeshore. As we pulled out of the bay, I spotted another winged hunter—a bald eagle perched on a high, lone pine. Even from a distance, the adult was an impressive-looking bird. The eagle stood almost one metre high and its hooked, yellow beak and piercing, yellow eyes gave it an aggressive appearance. The eagle's snow-white head and tail feathers contrasted boldly with its dark brown, almost black body.

Unlike northern harriers, who are on the wing most of the day, eagles spend their time on favourite perches unless disturbed or stimulated to action by the sight of prey. Immobile except for its head, which slowly swivelled from side to side as if moved by the breeze, this bird seemed almost a part of the scraggy tree. Eagles can spot mice or rabbits moving in the grass from more than a kilometre away—even though this bird appeared to be resting, we knew it was searching for prey.

After almost an hour of scanning, the bald eagle spread its broad wings, which spanned more than two metres, and floated off its high perch. Instead of dropping straight down as the harrier had, this bird descended in a

Harrier and Eagle

long, gliding trajectory and cruised low over the lake. With wings outstretched and black talons extended, it skimmed the surface until it had snatched its prey from the lake.

The fish must have been a large one. The eagle flapped its wings on the surface of the water and lunged upwards several times as it tried in vain to rise from the lake. An eagle is a powerful bird and can lift a weight equal to its own, but this fish was too heavy to allow the bird to become airborne. Finally, the hunter resorted to towing its victim ashore, swimming in a most ungainly manner by using its long wings as paddles on the water's surface.

It was a lengthy and tiresome process. When the eagle finally reached the edge of the lake, it hopped on one foot, dragging the huge fish onto the beach with the other. Before tearing small bites from the fish with its hooked bill, the eagle paused briefly to preen its feathers—probably to remove the water. To us, however, the eagle seemed to be grooming itself for a fine meal.

After feeding for several minutes, the eagle threw its head back and let out a series of sharp cries, "ki-ki-ki-ki-ki-ker." The voice was surprisingly weak, which seemed out of keeping with such an impressive-looking bird. The cry was almost like a gull's, but was broken into a succession of seven or eight staccato notes.

At this call, two immature eagles arrived to take turns at the feast. The young birds were a drab brown, with some specks of white showing on their chest. These two eagles wouldn't develop the distinctive white head and tail coloration of an adult until their fourth year. Because the adult bird was still feeding them, we speculated that they had been born that spring.

The first juvenile to reach the fish protected its prize by mantling over it with outstretched wings. The other complained vigorously, raising its beak skyward and screeching. Between these noisy displays, the second juvenile tried to snatch the fish from under the other's wings. The

A majestic bald eagle

harassing tactics worked, and the first juvenile finally left the remains to its noisy sibling.

Even though the bulk of an eagle's diet during the summer is fish, the bald eagle is more a scavenger than a hunter. Meals are often made of dead fish found on the shore or half-dead spawning salmon plucked easily from the rivers. Bald eagles also feed on ducks and other waterfowl, especially during the winter and spring when fish are more difficult to find. Given the chance, they will pirate prey from other raptors. Sometimes an eagle will harass an osprey on the wing until it drops its fish, which the eagle will then catch deftly in mid-air.

Eventually, the adult departed from the lake. The large bird's flight was powerful and steady as it moved with slow beats of its great wings. The eagle flew over the marsh, flushing up flocks of ducks ahead of its arrival. No duck was going to be caught on the water with an eagle for company, and string after string of mallards, widgeons and small teals took to the air and sped away.

The eagle continued along the marsh, then soared upward, carried aloft by the mid-afternoon thermals. With each turn, sunlight glinted on the bird's white head and tail. The eagle, with its slow and ponderous wingbeats, was less graceful than the harrier we had seen earlier. But bald eagles are masters at soaring upwards on wind currents, assisted by their wide wings and broad tails.

Now the eagle circled higher and higher, riding the wind currents until it seemed only a speck in the blue. Sally and I sat there, earthbound, and marvelled at the freedom of this magnificent bird—an untamed, untouched creature of the clouds and sky.

A bull elk—bugler of the forest, pages 11–20

66 TOP: The rut—bull elk sparring, pages 11–20
BOTTOM: A burly black bear, pages 30–37

A curious caribou, pages 45–52

68 TOP: *A friendly chickadee, pages 129–135*
BOTTOM: *A ruffed grouse "budding," pages 38–44*

A Canada goose on her nest, pages 105–113

A least chipmunk, pages 153–160

TOP: A pika announcing our arrival, pages 53–57
BOTTOM: A prickly porcupine, pages 21–29

A bull moose coming up for air, pages 170–178

A mule deer in winter, pages 114–120

A hoary marmot basking in the sun, pages 121–128

A coyote on the prowl, pages 126–128

A mountain sheep at a lofty lookout, pages 145–152

TOP: A marten hunting for mice, pages 136–144
BOTTOM: A beaver eating aspen leaves, pages 161-169

78   TOP: *A sassy red squirrel, pages 98–104*
BOTTOM: *An arctic ground squirrel, pages 179–182*

A young elk's first winter, pages 11–20

A goat shedding its winter fur, pages 81–89

NINE

Goat Country

Because of the high country they call home, mountain goats are among the most difficult creatures to photograph. In fact, our misadventures began before we even saw one of the woolly white animals. Sally and I had been following goat trails across crags and cliffs for two days, and rather than drop back to the valley each night, we pitched our tent on the only piece of flat terrain on the mountainside. The narrow ledge had seemed quite adequate... until the wind came up in the middle of our second night.

The strong wind seemed determined to tear the tent pegs from the ground. As the wind gusted higher and higher, I thought that we would be blown—tent, sleeping bags, clothing, packs and all—to the valley floor, almost six hundred metres below. We lay awake listening to the howling wind and cringing as each stentorian blast shook the tent. At the first hint of morning light, we crawled out of the tent.

"This place is fit only for goats," I grumbled as I stuffed

the tent into my pack.

"...and photographers crazy enough to climb up here!" Sally replied.

Feeling tired from our sleepless night and discouraged at not having photographed any goats, we shouldered our backpacks and began to search for a sheltered spot on the windswept mountain to cook breakfast. We hiked around to the leeward side of the mountain and there, only one hundred metres away, stood a nanny goat and kid—as though they had been waiting for us.

Mountain goats have incredibly keen eyesight. On previous days we had found that by the time we had spotted a herd way off in the distance, the herd had already spotted us and was moving away. Because we had never been this close to goats, our breakfast was forgotten as we hauled cameras and tripods from the packs.

"Be careful... it's a couple of hundred metres straight down to the next stop," I warned Sally as we clambered down the loose rock. But calculated risks are half the thrill of wildlife stalking. We moved closer, then stopped for a moment and watched the wind playing in the nanny goat's long white fur.

Suddenly, a rock gave way underfoot and I began sliding down the slope, rocks clattering under my boots and crashing down the mountain. I backpedalled as I fought for a foothold—anything solid for my boots to get a hold on—and finally came to a halt a few metres later, precariously perched on an outcrop of granite.

"Is the camera equipment okay?" Sally called down. This was a rather overused joke, but served to break the tension of the moment. With a thumbs-up sign I indicated I was unhurt, then paused to catch my breath and ponder the sanity of climbing across steep mountainsides, over unstable scree slopes and to dizzying heights in pursuit of these elusive animals. I had gained a few bruises and a gash on one hand from trying to grab rocks on my way down, but was otherwise unscathed.

A nanny goat and kid on a high lookout

Mountain Goat

When I stopped trembling, I turned to look for the goats and was surprised to find that they hadn't moved; the rumbling of rockfall must be an everyday sound in the mountains. In any case, they probably figured, after seeing my act, that they had nothing to fear from me. The nanny goat stood with her four feet planted firmly on a boulder, seemingly oblivious of the fact she was on the edge of a dropoff. Goats have short, stocky legs with large hooves which are specially designed to grip rocks. Each hoof has a concave, cushioned pad that helps them to climb the steep terrain with ease.

The nanny's jet-black hooves, horns, eyes and nose contrasted sharply with her shaggy white coat. Her fur was thick, with long guard hairs over a dense mat of fine wool, but the early summer moult had begun and her coat was ragged. Fur was falling out in tattered clumps, revealing a soft-looking new growth of fuzz over her gaunt body. I had visualized the goats as large-framed animals, but most of the bulk was actually fur. I could now see that the mountain goat was slim, slab-sided and narrow shouldered so that she could manoeuvre easily along narrow mountain ledges.

Sally and I walked closer, carefully picking our way across the scree slope to follow the nanny and kid as they moved to join a group of other goats. The goats walked just as slowly, choosing their footing deliberately and with care. Survival in this precipitous terrain, as we quickly learned, comes from being cautious.

As we approached the herd, we could see that it was made up of a number of nannies, kids and yearlings. Goats band together in small groups of three or four families and sometimes in herds of up to thirty animals. There were no billy goats to be seen. They play no part in raising the family and remain far from the herd for most of the year.

The half-moulted nanny goats were not very attractive, but their young kids were the most endearing creatures

Goat Country

we had ever seen. They were only a month old and had short, curly hair and round baby faces. They were fluffy white balls on stocky legs and hid behind their mothers like shy children, peering at us with black, glistening eyes. We heard few sounds from the goats except for the occasional high-pitched "b-a-a-a" from a kid that had become separated from its mother.

Sally and I came to know the goats better during the following days. In early morning they usually moved down to feed on the sparse grass of the high slopes. The goats spread out across the slope and would not tolerate any other goat near their grazing area. They signalled their displeasure by stamping their front feet and flicking out their tongues. When one young goat ignored these signs and approached too close to a nanny and kid, it was charged by the nanny, who appeared ready to use her sharp horns to protect her clump of grass and youngster.

I had often wondered how goats could possibly find enough food on the high, treeless mountains to survive. Even during winter they remain on the mountains, feeding wherever the wind sweeps the slopes bare of snow. Their answer to the limited supply of vegetation is to consume anything that looks even remotely edible: grass, moss, lichen, twigs, and even shrubs and stunted sub-alpine fir.

As with most things the goats did, they grazed at a leisurely pace. After feeding they wandered slowly upslope to a high ridge pockmarked with resting hollows. Each animal bedded down in a hollow to chew its cud; many seemed to have a ritual of pawing at their hollow to loosen the soil before lying down and rolling in the dust. One goat, already ensconced in her favourite hollow, pawed dirt onto her body with a forefoot, then scraped some more and sent a spray of dust over her back. This dusting assists the moulting process and also helps rid their bodies of fleas and ticks.

The goats paid little attention to us as they settled into

their beds; they must have known that they were in a place we couldn't get to. Few predators bother the goats in this high country, where only the occasional eagle or mountain lion will dare try for a kid. The goats are in danger from bears and wolves only when they graze on the lower slopes or drop even further to the mineral licks in the valleys.

Unfortunately, the goats had so much confidence in their lofty lookouts that they were all positioned with their rears towards us. Because back ends of animals make less than spectacular photographs, we discussed ways of attracting their attention. I clapped my hands, but this caused only one goat to turn its head, look at us for a moment, then turn away again. A shout brought a similar lack of response. In a final attempt to attract their attention, I began singing a rather off-key rendition of a ragtime tune.

This seemed to work. They stood up and gazed at this unfamiliar assault on their senses. Sally took a number of photographs of the animals as they studied us, unblinking and stately, unfazed by my strange performance. But the goats soon tired of my show and settled into their hillside hollows again. Later that day they moved even higher to the safety of the inaccessible crags and cliffs.

After a few days of clambering up, over and around the steep mountainside, we discovered that the best place from which to watch the goats was near a natural mineral lick. Sally and I were up early one morning and set up our tripods and cameras where we had a clear view of the path the goats would take. Our plan worked. Within a short time we were approached by two nannies who stopped to study us, their kids huddled tightly against their sides.

All the goats we met had displayed a strange mix of shyness and curiosity, and these two were no different. Both nannies stared at us for several minutes, their heads tilted to one side, as if trying to determine why we had in-

Mountain Goat

truded into their domain. Finally, the nannies moved cautiously towards the lick.

The animals became so intent upon satisfying their early-summer craving for minerals that they took little notice of us. Several other nannies and kids moved in, crowding each other for the small pockets of mineral-rich clay that had been worn hollow by many seasons of licking. We watched them lick and paw at the area, and were surprised to see several goats actually eating the clay. Occasionally, a goat would be crowded out as the animals shouldered each other to reach a choice spot, but they were far more tolerant of each other here than they had been in their grazing areas.

After a week of photographing the shaggy, shedding goats, we decided to return in the fall when the nannies would be more photogenic in their full winter coats. By then the billies would be with the herd, adding the excitement of their rutting activities to the scene.

When Sally and I returned to the mountainside in early November, we discovered that the goats no longer frequented the lick. After searching the area for half a day, we located the high-country dwellers grazing on the other side of the mountain. As we expected, the billies had now joined the herd. We could distinguish them from the nannies by their larger size and longer horns. All of the goats had long, white beards and had grown regal coats in preparation for winter on the windswept ridges. With their high, bulky shoulders and thick pantaloons of wool that ended at their knees, the goats appeared to be top heavy.

The billy goats moved with deliberate, stiff-legged steps, as if they were parading for the females; this was the beginning of the rutting season, which would last until December. The mountain goat's mating rituals are less spectacular than those of most hooved animals. Goats prefer bluffing to battle because the use of their sharp, stiletto horns could result in serious injury to both winner and loser. Or maybe they don't like to run the risk of being butted off a ridge!

We watched two billies rush at each other, viciously raking the air with their dagger-like horns, but always stopping short of actual contact. Slowly, they circled each other, arching their backs and raising the crest of hair on their necks and backs, clearly trying to make themselves appear larger and more impressive. When the larger of the two goats pawed the earth with a front hoof, the other billy decided this was his cue to leave before any real fighting could begin.

Even after the confrontation the victorious male was not assured of the right to mate with the female. Although the billy was larger, the nanny goat's horns were just as sharp and dangerous, enabling her to decide if and when the event would take place. The male moved slowly towards the female, looking for signs of aggression or acceptance. The female stood absolutely still, as if mulling it over, then slowly turned and walked down the goat path, the billy following cautiously behind.

Sally and I had been so absorbed in the drama of the high-country confrontation that we hadn't noticed dark clouds scudding over the crest of the mountain. When a howling wind almost blew over a tripod we reluctantly turned to leave. As we headed down the mountain, the first wind-driven snow flakes of the season drove against our exposed faces. Sally and I paused part way down the ridge and looked back up toward the goats; all we could see were two white, ethereal shapes in the snowstorm. The animals stood stoically on the high ridge, their thick coats protecting them from the elements.

Winter and summer, this windswept, barren country was the domain of the goats. These hardy mountaineers would survive the snowstorms and sub-zero temperatures... but it was no place for us.

See colour photograph on page 80.

TEN

Call of the Wild

As the winter sun set, the silence of the northern forest settled over our camp like an invisible blanket. Sally and I drifted off to sleep in our small tent, pleasantly tired from a long day of backcountry skiing. A few hours later I was drawn from a deep sleep by a sound. Drowsily, I wondered what had wakened me. Then the sound came again.

"Ow, ow, owwhoo-oo-oo." The wavering howl of a timber wolf floated on the still night air. It was a chilling sound that sent shivers down my spine, but it was also eerily beautiful.

Two more wolves joined in, creating a harmony of howls, one baritone voice maintaining the low notes while the others started higher, then descended in tone and volume. Through the thin walls of the tent, I could hear each wolf clearly as the notes swelled and faded.

"Sally," I whispered, nudging her through the sleeping bags.

"I hear them," she replied. "They sound so close... like

they're right beside us!"

Sally and I listened intently as the wolf song echoed in the night. We lay still and silent, alone with our thoughts. I knew that wolves had never attacked people, but it was difficult not to dwell on images of mythical wolves with fierce, flaming eyes and flashing fangs.

The wild calls seemed near, then far away, then near again as they wafted through the forest. We wondered where the wolves really were. The next morning Sally and I went searching for their tracks, eager to see how close to our camp the wolves had come. They had seemed to be standing almost at our tent's door, and we were surprised there were no tracks encircling our camp. We had to ski for five minutes before we saw the first sign of the animals.

The wolf tracks were easy to follow and left quite a story in the snow. During winter, wolves hunt in packs of six to twelve animals, usually all members of the same family. But it was difficult to determine how many were in this pack because they had walked single-file. The snow was deep and soft, making it too strenuous for a wolf to make its own trail for long, so they had stepped directly into each other's tracks. It must have been difficult breaking trail, because their bodies left a deep trough in the snow. Now and then, the lead wolf had stepped aside to change position with another.

Every change of pace, every detour, told a story of the wolves' activities during the previous night. Occasionally, a set of tracks veered off the trail where a wolf had investigated something or sprinkled an exposed log or boulder to mark the pack's territory. Each time, the wolf had rejoined the others further on. When the wolves had reached a small clearing, their tracks scattered in all directions. I imagined the animals romping in the snow, dark silhouettes bounding through the clearing, sending up sprays of snow in the silver moonlight.

The tracks converged further along where the wolves

Call of the Wild

must have held a "meeting." An area three metres in diametre was packed down and we could see the shape of wolf bodies where they had sat or lain in the snow. The glazed impressions and strands of hair frozen to the snow indicated they had remained for a long time.

The wolves had continued along a river where travel was easier on the firm snow. The tracks fanned out as each wolf investigated the many scents along the way. Now we could clearly see every separate paw mark—they were the size of my open hand, each depression showing the imprint of four claws.

Around the next bend the tracks converged upon those of a moose. We noticed a change of pace; the wolves had fanned out to four abreast and we could tell by the length of their strides that they had broken into a run. Like most wolf hunts, this was a team effort with the pack split up so they could work in relays to tire their victim. Two sets of wolf tracks veered sharply to the right, then splayed-out hoofprints indicated the moose was running at full speed from the pack.

The plot became more intriguing as we followed the tracks to an open meadow. It developed into a tempestuous story of survival in the wilderness as we found slurred tracks, patches of moose fur and bright red spatters of blood on the snow. We saw signs of a scuffle, then the tracks turned into the forest. Our heartbeats quickened as we speculated about the fate of the moose. We continued slowly, not sure if we really wanted to know the outcome.

The wolves had followed the moose only a short distance into the forest, then veered away from the moose tracks and slowed to a trot. It looked as though the moose, with its long legs, had been able to escape from the wolves, who had floundered in the deep snow. Wolf packs are most successful when they can chase an animal into an open area or onto the ice of a lake. But even then, a healthy moose can outrun wolves or defend itself by lashing out with its sharp front hooves.

Wolves are not the ruthless killers portrayed in many stories and fables. The much-maligned wolf is in reality just another wild creature earning its keep the hard way. They are certainly predators, and hunt moose, caribou, elk and deer through the winter, but they prey mostly on weak or sick animals. Wolves help maintain nature's balance, and by culling the sick animals, they actually benefit the prey populations.

Individual wolves are not physically formidable, and for most of the year their diet consists of smaller prey, such as beavers, hares, ground squirrels and mice. Wolves stand considerably less than a metre high at the shoulders, weigh much less than the average adult human, and have neither great speed nor stamina. They must work hard for their meals; they fail in their hunts far more often than they succeed.

That winter, Sally and I often heard wolf serenades and found many tracks, but the wolves were shy, elusive and generally nocturnal. Many days we skied through their territory, hoping to catch a glimpse of the animals, but we were usually unsuccessful. Our luck changed early one morning when we saw a grey shadow, then another, drift in and out of the trees on the shoreline across the lake. Then, from the evergreen cover, a large wolf came out onto the lake, followed by the rest of the pack.

Breathlessly, we watched as the wolves, their muzzles raised, searched for a scent, any scent. Wolves have sharp eyesight and keen senses of hearing and smell, but it was snowing heavily and the wolves had not yet noticed us. Through a telephoto lens, I could see them moving along the lakeshore at a fluid trot. They moved with seemingly effortless motion, as if blown by the breeze rather than powered by muscle.

When the lead wolf detoured to a rock and lifted a leg, we realized he was a male. He was a striking creature, primarily grey with a sprinkling of brown along his sides

A timber wolf in a snowstorm

and a black-fringed mantle of fur over his shoulders. His head was broad, with a long, rounded muzzle and prominent ears. His long, thickly furred tail was carried straight out behind, a sign that he was the dominant wolf.

The other wolves were slightly smaller; we immediately noticed that they held their tails loosely, and lower, than did the lead wolf. Although northern timber wolves tend to be mostly grey, the wolves of this pack ranged from shades of grey-brown to near-white.

The lead wolf was treated with respect by the pack members. Whenever a wolf approached him, the visitor's ears would be laid back and its tail tucked between its legs. Occasionally, a wolf would drop onto its front knees and crawl to him—and we even saw one wolf roll over onto its back in submission. The alpha wolf remained aloof and seemed to ignore most of this behaviour at first. When he finally wagged his tail in an expression of good will, the other wolves responded playfully and licked or nipped at his face and neck.

Wolf packs have a highly organized social structure, which centres on the dominant male and female. These pack leaders usually mate for life and are generally the only breeding pair in a pack. However, all members of the pack help with the raising and feeding of the pups, and also teach them hunting and survival skills.

The pack began trotting away from us and, hoping to attract their attention, I gave my best wolf howl. It took a second or two for the sound to reach them, but when it did the wolves came to a skidding halt. They turned in unison and stared across to where we stood. The wolves were only fooled once though; my second howl caused them to disappear into the forest like wind-blown smoke.

We saw the wolves only a few times after that encounter, since the pack was constantly on the move, even in the deep snow. Wolf packs have a territory that can cover up to a hundred square kilometres. They may roam

Call of the Wild

more than sixty kilometres a day in search of food, and home is wherever they happen to lie down to sleep. Wolves live in dens only during the spring and summer when the young are born and raised.

Whenever the wolves were in our area, however, their mournful medley echoed in the still night air. Howling is said to be a way of keeping in touch with pack members, or a way of warning other packs away from occupied territory. I like to think that it might be a form of "songfest," for wolves apparently enjoy a good howl.

Sally and I tried several times to join in their song. Although the wolves seemed to enjoy Sally's soprano howl, they often stopped at the sound of my deeper call. This, of course, was not flattering to my ego.

I will always remember one special evening later that winter when Sally and I howled together and were answered by the call of wolves. We listened with delight as the valley became a concert hall filled with wolf music. As we sat around the glowing campfire, howling into the darkness, we felt a special bond with the wolves; by answering our call the pack seemed to accept us as part of the wild country they roamed.

"Ow, ow, owhhoo-oo-oo." No other sound could ever evoke so well the feeling of wildness, vastness and seclusion we experienced in the northern wilderness.

ELEVEN

The Sassy Squirrel

Of all the creatures we met during our travels in the northern forest, the most spirited and vocal were the red squirrels. Although only diminutive balls of red fur, they brashly scold any and all intruders into their domain with a loud, chattering "tchrrrrr!"

We found that it was almost impossible to walk through the woods unnoticed. Our progress was always announced by a succession of squirrels who chattered, barked and scolded from their treetop posts: a wilderness telegraph system that informed every inhabitant of the forest about our presence.

The noisiest of all was a red squirrel who lived near our cabin. The first time I walked through his territory the squirrel startled me with a loud churring from a branch just above my head. He was a chatterbox who seemed to delight in the sound of his own voice. The sassy squirrel held one paw at his chest and stamped his tiny, furred feet, scolding me with his abusive vocabulary. Each time he let out a ratchet-like churr, his rusty-red tail stretched

The Sassy Squirrel

backwards and flicked up and down as though attached to his vocal cords.

"Now Rusty, what's all the fuss about?" I asked in a low voice, trying to calm the agitated animal.

Rusty replied with another string of expletives. I was amazed that such a small creature could be so aggressive and loud. Then Rusty turned and scampered up to the safety of the highest branches. There he told off any onlookers with a ringing call that echoed through the forest.

It seemed that I had interrupted the squirrel during his busiest time of year—it was late August and he was gathering cones for the winter. Rusty leapt from branch to branch, quickly cutting off cones and tossing each one down with a flick of his head.

One cone landed squarely on my head, prompting me to step back a few paces as the edible projectiles rained down, rattling and bouncing noisily among the branches. After several minutes of frenzied acitivity, Rusty descended and began collecting his harvest. Wasting no time, he picked up a cone in his mouth and ran to the base of the tree, where he dug a little hole, pushed in the cone pointed end first, and covered it.

Rusty stored the cones in many scattered locations and in at least one midden—a large pile of cones, cone scales and other debris. His midden was almost two metres across and as high as my knees. In fact, the pile was so large that many generations of squirrels must have used the same midden.

The squirrel returned again and again to bury his winter's food, one cone at a time. After almost half an hour of nonstop work, Rusty carried a cone to a low branch and began eating. Even while eating, his pace didn't appear to slow down. He nipped off the scales quickly, turning the cone round and round with his deft, long "fingers." When he reached the stem, he discarded the core and descended the tree for another.

I picked up a cone at my feet and inspected it; when I

The Sassy Squirrel

peeled back one of the many scales I found a pair of small seeds tucked underneath. These tiny morsels of food were what the squirrel had been working for so industriously. Red squirrels will eat almost anything, but dine mostly on nuts, cones and mushrooms. In our travels, Sally and I had often spotted large mushrooms on the outermost branches of spruce trees where a squirrel had placed them to dry in the sun before being stored.

After his meal of cones Rusty paused to clean his spruce-gummed whiskers and paws. Then he groomed the rest of his body by wiping his paws across his fur, paying special attention to his tail, which he cleaned, combed and fluffed with claws and teeth. I almost laughed at his contortions; he twisted backwards, straining to reach an inaccessible spot on his back, then lost his balance and nearly fell off the branch.

The squirrel seemed to have chosen a sheltered place to complete his grooming, but even then he froze in position when a raven passed over, and he looked around when another squirrel scolded. These small rodents have to be constantly on the alert for predators—owls and hawks from the air, fox and lynx on the ground, pine martens in the trees.

In addition to predators, red squirrels have to be on the lookout for other squirrels intent on raiding their middens. A squirrel's home range might be only two hundred metres across, but it is three dimensional; the forest floor is as much a part of their range as the trees.

A few days later Sally and I noticed another squirrel approach one of Rusty's middens. We anticipated a confrontation—squirrels are solitary creatures and do not tolerate others in their domain except during the brief mating season of late winter. As we'd expected, Rusty began scolding loudly in a tirade of expletives. "Tsch-r-r-rrr, siew, siew, siew."

"Doesn't sound like a friendly greeting," Sally observed.

Red Squirrel

Seconds later, Rusty dashed down the tree towards the intruder. The two squirrels raced up and down, round and round the tree trunk, across a stretch of forest, then up another tree in a blur of red. The forest echoed with their loud scolding notes and the scratching and scraping of their claws on the bark.

They bounded at high speeds, leaping from branch to branch at dizzy heights. It looked as though Rusty had almost caught up with the other squirrel when it propelled itself in a flying leap across two or three metres of empty space. We held our breaths in suspense until the squirrel landed safely on the swaying, springy branch of a neighbouring tree. Rusty also jumped, but missed the tree and fell, with legs and tail outstretched, almost six metres to the ground. Fortunately, his large tail was a parachute of sorts, and Rusty seemed unharmed; the mad chase continued until the intruder was finally evicted.

Rusty hurled one last "tchrrrrrr" after the fleeing squirrel, then resumed gathering cones as if little had happened.

As the cooler weather of autumn settled in, Sally and I often saw Rusty, still gathering food in the forest. His territory bordered on the clearing near our cabin and he regularly scampered to our bird feeder to steal grain meant for our feathered friends.

We were not pleased with Rusty raiding our bird feeder so often, and tried many different feeders in an effort to thwart him. He easily mastered each one. I thought I had finally outsmarted him by nailing an old gold pan to the top of a tall pole—there seemed no way he could scale the pan's smooth sides. We were surprised the next morning to see him sitting comfortably in the pan, munching on birdseed. He ignored the plaintive calls of the chickadees and the screeching of the jays while he helped himself to the choice offerings.

Finally, we strung a wire from our cabin to a tree and suspended a metal camping plate from the wire. After

"Rusty" on a safe perch

Red Squirrel

one attempt at walking along the wire, Rusty gave up. He vented his frustration by scolding nonstop for almost an hour. Every bird stayed away during this tirade. A few days later it seemed a suitable arrangement had been worked out: each time a bird fed in the tippy feeder, enough grain spilled onto the ground to appease Rusty.

When the first snowfall blanketed the forest, Rusty stopped coming to the cabin. Red squirrels do not hibernate, but they do become less active during cold weather. On the coldest days they remain in their dens, either deep in their middens or in tree hollows. We had often seen Rusty pop out of an old woodpecker's hole in a nearby tree; he had gnawed the hole wider with his tiny teeth and we guessed this was his home for the winter.

During one week of extremely cold weather the forest was strangely quiet.

"I almost miss Rusty's call," Sally commented as we snowshoed through the forest. We could imagine Rusty in his den, nestled in a bed of dry grasses and protected from the icy cold. Unlike us, squirrels seemed to have the sense to stay indoors on very cold or stormy days!

As soon as the cold spell ended, the forest was again dotted with tracks and tunnels as the squirrels burrowed out from their dens in search of stored seeds and cones. Although the cones had been hoarded months earlier in many locations, small piles of scales near the tunnels indicated that the squirrels had no trouble rediscovering their caches.

We could usually count on seeing our furred neighbour near his den on the warmer days. One bright February day we found him perched on the outer branches of a spruce tree, sunbathing in the meagre warmth of the low sun. Even though the small animal was almost rolled into a ball to conserve energy, he was still active and sassy enough to hurl a series of chattering comments down to us as we snowshoed past.

"Same to you, Rusty!" I answered with a smile.

See colour photograph on page 78.

TWELVE

Wedge of Wings

More than any other event, the quacking, cackling, whistling tide of birds from warmer regions signalled the arrival of spring in the north. After a cold season of few subjects to photograph, mid-April suddenly brought more winged visitors than Sally and I could point a camera at. These birds added colour, sound and excitement to a land that was still asleep under a blanket of snow.

A week after the first mallards and widgeons had arrived, we awoke to hear the clarion call of another flock of visitors.

"Geese!" I hollered to Sally. We rushed out of our cabin and looked skyward, quickly spotting a long wedge of wings heading north. Sally and I watched the flock of Canada geese fly over the lake, their long, black necks outstretched and large wings undulating effortlessly. At the head of the V-formation was a solitary goose, taking its turn at the difficult task of leading so others would have an easier flight in the slipstream.

Canada Goose

For the next week more northbound honkers flew over our valley night and day, guided by some internal navigation system that directed them to the same nesting grounds year after year. Some of the flocks weaved in formation over the snow-covered valley, noisily honking back and forth as they inspected each newly melted puddle and pool, but none of the birds landed. We began to feel left out of the excitement as the wild flocks continued northward.

Then one morning the musical, honking medley became louder than usual as a flock circled low over the valley. We watched with anticipation as the large birds set their wings and glided silently downward in a long approach.

Just above the open water, the geese stretched their webbed feet forward and flapped their wings to slow their descent. Compared to their elegant flight, their landing looked awkward, with black feet and tails skidding across the water. Several landed on the ice, running clumsily to a stop. As soon as they touched down our newly arrived neighbours began honking and splashing, seeming to proclaim that winter was finally over!

The majestic geese had black heads and necks, with striking white chinstraps stretching from ear to ear. The rest of their bodies were shades of grey and brown, except for their white rumps and black tails. It was impossible to tell the males from females until they separated into nesting pairs because, except for differing sizes, all the birds looked the same.

These geese honked a clamorous invitation to each subsequent flock that flew over the valley, and by late afternoon the ice-free areas were crowded with vocal visitors. As daylight dwindled the geese became less boisterous. All of a sudden, a loud "haronk" broke the silence and in an instant the entire flock was calling and cackling. We heard the beating of wings on water and a general sound of argument, then a few birds rose up from the pond and circled before gliding back to their resting places. Finally, a deep-voiced honker had the last word and the noise again diminished to a scarcely audible banter.

There was still a metre of snow in the bush and few open areas of water. While the ducks and geese waited for the snow and ice to clear from the waterways, they flew back and forth between open spots along the rivers, paddled in the small areas of open water, and rested on the ice-covered lake. Each day more flocks arrived, but most stayed only a few days, then continued further north.

Canada geese mate for life and tend to return to the same breeding grounds each year; we were delighted to see that several families had decided to stay in our valley. Towards the end of May when the ponds and then the lake had finally broken up, the birds became difficult to locate as they dispersed to the marshy areas at the water's edge.

It was a challenge to photograph the wary creatures. Each small group of geese was guarded by one or more sentinels, who were constantly on the alert. Their eyesight is very keen and their sense of hearing acute, making them among the most difficult birds to approach. At a warning note from a watchful sentry every head was raised and eyes fixed in our direction. We moved a little closer, causing them to lift their chins and shake their heads, indicating they planned to take flight. In a cacophony of splashing, flapping and honking, they took off and circled several times before landing in a new location.

After several days we discovered that the geese were quite regular in their feeding habits. The flock returned day after day to graze in the same meadow and we often watched them feeding early in the morning. The geese became more wary during mid-day when they rested, but by late afternoon they again became preoccupied with dining on the grasses and dabbling in the shallows.

Wedge of Wings

In early June Sally and I watched with amusement as two mature adults tried to drive away their previous year's offspring. The young birds were reluctant to leave—they had been near their parents since hatching, had flown south in the autumn with them, wintered together, and made the return flight north together. But the time had come for the adults to begin nesting and to raise another family; they no longer wanted the yearlings with them.

From our hiding place we watched the gander try to drive the young away again and again. Each time the parents swam off in search of privacy, the young birds swam after them, determined to stay close. Finally, the gander lifted from the water and charged at the yearlings, honking angrily, and sent them running across the water.

After circling twice the gander dropped down onto the water to join his mate. By the time the gander had rearranged and settled his feathers, the young were right back where they started and the honking chase began again.

"I wonder if they will ever get the hint," Sally mused after half an hour.

It took several days for the young geese to resign themselves to the situation, but finally they left the lake. The older geese were now free to build a nest and raise another family.

Sally and I decided that we should leave the two birds alone for the next few weeks. They would need time to build a simple nest of reeds near the water, and the incubation period for their eggs would last almost a month. Although ganders are staunch defenders of their nests, the eggs are vulnerable to predators during this time. Foxes, coyotes, ravens and crows are among their enemies, and we didn't want our presence to add to the birds' stress.

When we returned to the nesting area in July, the gan-

Canada Goose

der swam out to intercept us. All the wildness and wariness seemed to have gone as he looked at us steadily and quietly. We let the canoe drift closer while studying his reaction—it was as if he was deliberately luring us towards him. We quietly beached the canoe and sat down on the shore while the gander swam back and forth on a short beat in front of us. As the minutes passed and we made no hostile sound or movement, the gander swam away and began to utter a curious, low gabble.

Out of the grass, only four canoe-lengths away, a head and long dark neck rose to look at us. From somewhere nearby we heard low cheeping and whistling sounds where several goslings had been hiding in silence.

I got up from where we were sitting and edged towards the young family. But as I approached, the gander rushed towards me and stood in my way, shaking his head defiantly as he honked, ready to defend his mate and offspring. I took another step and the large bird ruffled all his feathers and spread his wings until they reached almost two metres across. He looked fearsome enough to scare away most prowlers. Another step and he lowered his head and black neck close to the ground and rushed straight at me, hissing loudly.

I took two quick steps backward. The gander stopped, his head close to the ground, mouth open and pink tongue curled as he hissed vigorously.

"Those wings could cause a lot of damage," I whispered to Sally when I sat down. We were impressed at the way the goose had been prepared to tangle with an intruder many times his size. At least I had *thought* I was many times his size until I saw his wings fully extended!

We sat motionless for a long time, until the gander must have decided we meant no harm. He uttered a low call and his mate slipped out of her hiding place onto the water. Following her were six goslings, alert little bundles of yellow-brown fuzz, who paddled madly to keep up. The gander continued to watch us as the family

A goose watching over her goslings

Canada Goose

glided into the reeds. When they were out of sight again, he settled into the water and swam over to join them.

On our next visit we set up a blind near a stretch of muddy shoreline that was laced with goose tracks. From our vantage point we watched the group clamber onto shore and begin feeding. Because their feet are further forward than those of ducks, geese walk easily on land. Although we had seen them upending in the shallows to feed on underwater plants, these birds spent most of their time grazing on grass, clover and other vegetation.

The gander always hovered nearby, guarding his little flock. When they went for a swim, the goose usually led, with the six bobbing goslings strung out behind her and the gander acting as rear guard. The goslings occasionally strayed when they played games and chased each other across the water. Whenever one of the goslings became separated from the rest of the brood it gave a loud and frantic "peep, peep, peep." The female quickly swam to the rescue and pushed the youngster back into line.

The goslings grew rapidly and within two weeks had more than doubled in size. Their yellow feathers began to fade to a rather dirty grey, except for a few darker accents on the head and neck. They were gangly-looking creatures, with nothing to suggest that they would grow to be handsome Canada geese.

After six weeks adult colour patterns began to show, from white cheek patches to the black colouring on their heads and necks. Their downy fluff was replaced by flight feathers and by the end of August they were almost fully grown and closely resembled their parents. On several mornings we saw the young geese practising their running takeoffs, then winging from pond to pond, building up strength for their upcoming flight south.

As the autumn leaves turned colour, the gander and goose became restless. They stretched their wings, preened their flight feathers and fidgeted on the water. With each passing day their raucous calls became more

Wedge of Wings

vigorous and frequent. One frosty morning we heard a faint honking drift across the northern sky. Several families of geese rose from the marsh, honking loudly as they gained altitude to join the wavering V of geese heading south.

To us this flight seemed different, less certain than the spring migration; the lines of the flock were less defined and often broken. The clear full-throated calls of spring were replaced by a communicative chatter and the faltering, higher-toned notes of the young birds. For many of the geese, this was the first long flight away from the security of their birthplace to wintering grounds that might be as far as five thousand kilometres away.

Perhaps we found their departing cry less stirring because the two-syllable "haronk" of the geese signalled the end of their stay in the north. Each day more flocks passed overhead, some wedges of wings so high they were barely visible. By the time the first flakes of snow drifted earthward, we no longer heard the wild call of geese. We would have to wait until next spring to hear their exuberant voices again.

See colour photograph on page 69.

THIRTEEN
Mule Deer

The first rays of daylight filtered through the forest as birdsong welcomed the new day. It was a cold morning in late March, and after more than an hour of sitting motionless near a deer trail, I was almost too chilled and numb to endure my early-morning vigil any longer. I had seen two squirrels, three jays and a pine marten... but no deer. Just fifteen minutes more, I thought to myself. If I sat by the trail long enough, surely a deer would wander my way.

Finally, I heard the faint sound of an animal approaching across the crusted snow. I held my breath and studied the game trail intently. In the flat light I could just make out a form moving along the trail, pausing every few steps, then continuing towards me.

It was a mule deer, walking slowly and taking care to pause in the shadow of the trees, where it was well camouflaged. I immediately forgot about my cold hands and feet and the long time I had spent waiting. The animal stepped cautiously into the clearing, emerging

Mule Deer

from the shadows so silently that it appeared to be a part of the scenery. Then I saw several more deer moving down the trail. This did not surprise me—during winter deer are gregarious and the bucks, does and offspring travel together for safety. The more ears and noses to detect danger, the better.

The first deer was a doe, leading a small group of seven animals. She stood motionless for almost a minute, ears perked forward and black nose into the wind, listening and smelling for signs of danger. Finally, she began to nibble on willow twigs. As the others entered the clearing I noticed that one buck still carried his previous year's antlers. Another buck had recently shed his, exposing two blood-red pedicules on his forehead. Soon a new set of antlers would begin to develop, the small buds slowly growing to full size during the summer.

Each deer was pale-faced with a black forehead patch and a shiny black muzzle. I admired each deer's most prominent feature—an enormous pair of broad, mule-like ears, which give them their name. The deer were clothed in coarse, grey fur, and when one turned away from me I could clearly see its white rump patch and narrow black-tipped tail. Even though the females were smaller than the males it was difficult to distinguish a large female from a small male now that most bucks were without antlers.

Deer are delicate-looking creatures and their every movement while walking is slow and elegant. Their dainty feet are inadequate for pawing through the snow, so they depend on aspen and willow twigs, as well as the tender tops of coniferous trees poking through the snow for forage. These deer fanned out in the clearing, each searching for something to nibble on. One by one, they found their way to a fir tree that had fallen in a recent storm. This tree was a blessing for the deer, as it offered plentiful and accessible food. If the fir hadn't been there, the mule deer might have had to resort to browsing on

Mule Deer

standing trees, rising up on their hind legs to nip the less accessible branches. Deer often do this when food is scarce.

The deer continued feeding until the low sun crested the mountains, flooding the valley with light. Then they retreated into the protection of the forest to rest and chew their cud; they would remain hidden until the dim light of dusk again offered some protection.

I headed back to camp to warm up and share the excitement of my morning with Sally. We had tried several times that winter to get close to the deer, but they had always detected our presence, either with their strong sense of smell or their keen hearing. Deer are especially difficult to approach during winter when they are constantly tested by wolves and other predators. This had been a rare opportunity to observe a wintering herd from such close proximity.

The next time we encountered a mule deer was during late May, when Sally and I were resting beside a pond. A movement on the opposite shore caught our attention and we saw a doe walk slowly towards the water. The doe looked thin and weary after the long northern winter. Her summer coat of tawny-brown hardly covered her bony frame.

We watched patiently for several minutes while the doe drank from the lake. I tried to move closer with my camera but the faint crack of a twig under my foot was enough to startle her. She lifted her head, turned her large ears towards me and snorted with alarm. A deer depends on its lightning-fast reflexes to escape from danger, and in a second she had bounded away. The doe's movements were lively and graceful, all four hooves leaving and striking the ground together as she bounded along the shore of the pond. Her stiff-legged leaps must have been more than a metre high; she seemed literally to bounce off the earth as if powered by springs instead of muscle.

Mule Deer

At the edge of the clearing, the doe stopped to see if she was being followed. When we remained still, the deer took a few steps towards us, then bounded into the forest, as though trying to lure us away.

"She must have a fawn nearby," Sally whispered.

We searched the area for more than an hour and were ready to leave when I almost tripped over a well-camouflaged fawn, only ten paces from where we had first been. The fawn was lying motionless on the forest floor, its ears tucked flat against its head and its outstretched neck pressed to the ground. It was all but invisible. I knew that the fawn depended upon this illusion of not being there for its survival, as otherwise it would be defenseless. In fact, young deer licked clean by their mothers just after birth have almost no odour, making it difficult for predators to sniff them out. If there are twins, each is hidden in a separate location to increase the chance of survival.

The fawn was only as long as my arm, leading us to think that it was just a few days old. Its reddish-brown coat was dappled with countless creamy-white spots. The coat looked like a patch of sun-splashed brown earth. These spots would not fade until the fawn could easily outrun most predators, at about ten weeks of age. Then the young deer would no longer need the protection of camouflage.

I knelt down and spoke softly to the fawn. The young animal lifted its head and looked at me with enormous brown, trusting eyes. I longed to reach out and touch the fawn, to run my fingers through its soft-looking fur, to befriend it. But that would have taken its wildness away. By leaving my scent on the animal, I would taint its protective lack of odour and it would become vulnerable, lying there on the forest floor. Instead, Sally and I moved to a hiding place nearby, anticipating that the doe would come to feed her youngster.

A doe visits her young several times a day, stays only

An alert mule deer

long enough to feed it, then departs so that her scent will not attract predators to the resting place. When this fawn's mother finally appeared, she seemed to sense our presence. She was cautious and stood listening, looking, and sniffing the air. Finally, the doe walked to her fawn and reached her head down to check the faint scent of her offspring. She sniffed one fuzzy ear, then nudged the fawn to rise. At this, the young animal clumsily stood up and teetered after its mother, who led it further into the underbrush.

Sally and I chose not to follow them. Sometimes, sharing the wilderness with wildlife means letting the animals be. We had no photographs this time, but the memory of watching the most beautiful of woodland creatures was worth as much to us as any image on film.

See colour photograph on page 73.

FOURTEEN

Alpine Whistler

A short, shrill whistle pierced the morning silence and reverberated across the alpine slope. Sally and I stopped in our tracks as the call was repeated from the direction of a nearby rockpile—this was our welcome to a marmot colony in the mountains of northern British Columbia.

I scanned the area with binoculars, but although we could hear the marmot, it was difficult to locate the salt-and-pepper-coloured animal among the rocks. I finally spotted the furry creature sitting upright on a boulder, and further inspection revealed several other balls of fur in assorted poses on the slope. One lay on its back, pudgy stomach in the sun, one sat upright like the watchful sentinel we had first seen, and several others were sprawled out on their bellies.

Sally and I walked closer, prompting the sentinel to whistle again, this time with a more emphatic call.

"Eeeeeeeeeeeeee!"

The tranquil scene before us erupted into marmot-style

mayhem as the previously inert sunbathers scrambled over the boulders and disappeared, each one repeating the alarm as it fled. The brave sentinel remained atop a boulder, even after all of his companions had found hiding places. He repeated the warning call every few seconds as we came closer.

A marmot's warning can be heard up to a mile away. Not only marmots reacted to the call—we noticed that birds stopped singing and that several ground squirrels scampered for cover. On other occasions we had even seen mountain sheep raise their heads and look around for danger.

Sally and I sat down nearby and waited quietly. The sentinel remained at his post and we hoped the other marmots would come out and join him again. Sure enough, after five minutes or so one furry head, then another, poked out from between the boulders. They looked about tentatively at first, with their noses in the air, sniffing for the scent of any predator.

Like all hoary marmots, each had a black face with a white patch on its forehead and a white muzzle. Their thick coats were peppered with shades of brown, except for their feet and the tips of their tails, which were jet black. The species name, *caligata*, refers to their black feet; but a beautiful frosty-white mantle of fur across their shoulders and upper back makes the common name, hoary marmot, seem most appropriate.

Hunger eventually overcame the marmots' caution and they returned to their day's activities. It was early June and they had only recently emerged from eight months of hibernation. During the next four months they would have to do all of their breeding, playing, sunbathing—and eating. Their loose-fitting coats indicated they had a lot of eating to do!

We watched a marmot amble down the boulders to a meadow of lush green grass and flowering plants. The animal selected a purple lupine, nibbled it off at the base and munched it hungrily. The marmot chewed the flower as rapidly as a rabbit, using its deft, human-like "hands" to push the blossoms into its mouth. While feeding, the marmot was always alert and frequently stopped to scan every direction, including the sky, for signs of movement that might signal a predator.

The marmots were quite timid. Each time we tried to get closer they scrambled under the boulders through natural tunnels in the rockpile to their dens. The marmots had dug several burrows in the meadow, but they seemed to prefer the rockpile, which offered endless places to hide. We sat patiently for more than two hours. Sally finally suggested we return another day.

With each visit the marmots became more used to our presence and after several days we were able to observe their daily routine from a closer vantage point. Sally and I learned that a typical summer day in the life of the hoary marmots began long after sunrise when they lethargically emerged from their burrows. Each one would warm itself in the sun for fifteen minutes or so before beginning to feed. Eating was a long, drawn-out affair, often interrupted by socializing as they wandered about to greet other members of the colony. After a midday siesta on the warm boulders, the feeding resumed in the late afternoon.

A habit we found most peculiar was the way they flicked their bushy tails in the air before moving to a different rock or eating area. This eye-catching motion seemed to contradict the marmot's slow movements and good camouflage, but we concluded it was an effective way of keeping in visual contact with colony members.

The marmots had an entertaining way of greeting each other during their travels around the colony. Whenever two marmots crossed paths they would touch whiskers and each would sniff the sides of the other's head. During one greeting we watched one marmot give another a good-natured cuff that seemed to signal it was time for

Alpine Whistler

play. They both rose up on their hind legs and engaged in a shoving match, noses pointed skyward as each strained to push the other over backwards. When one toppled the other pounced on top, playfully chewing at its opponent's head and throat fur.

We were surprised to see the wrestlers suddenly stop for a moment and look around for danger. Seeing nothing to worry them, they continued their rigorous match, which involved tumbling and rolling together down to the bottom of the slope. After this tumble they broke apart, then ran back up the slope, one marmot chasing the other. Once again they tumbled down the slope, rolling and wrestling in a ball of fur.

Eventually, one marmot tired of the game and dove into a burrow. The other marmot pawed at the entrance but the retreat to the burrow seemed to signal the end of the match. After watching several matches, our conclusion was that the marmots enjoyed wrestling. They didn't appear to be fighting, and it seemed this was a way of working off aggressions among colony members.

Our favourite encounter with marmots was with a pair of youngsters later that summer. They were shy at first, and as we approached they turned and scampered into a burrow in the meadow. Fifteen minutes later, the young marmots appeared at the entrance of their burrow and began to play only a few metres away from us. They pulled, poked and stroked one another with their long deft fingers. The marmots tumbled and tussled, just like the adults we had watched, and were totally unconcerned about our presence. After a brief wrestling match they stretched out to sunbathe on the mound of dirt outside the burrow.

When they finally moved back to their tunnel Sally and I followed, crawling to the entrance and peering in. Soon one face, then the other, appeared at the opening, only a foot away from our cameras!

As we lay in the dirt I looked over to Sally. Gently, she

handed a blade of grass to a young marmot, who grasped it with tiny paws and began to nibble on the delicacy. While I took photographs, Sally was eye-to-eye with the marmot and smiling with contentment. The young marmot's acceptance of us meant they did not view us as intruders but as a part of their world. This trust, extended to us by these wild animals, was a highlight of our summer.

Over time the adults also became trusting and began to regard us as harmless creatures sharing their rockpile. One August afternoon we saw a marmot soaking up the sun on a huge boulder. Sally cautiously climbed onto the boulder with her sketch pad and began drawing the cooperative creature. The marmot was the perfect picture of relaxation, draped across the rock with one arm dangling over the side. The only sign of motion was its rythmic breathing. It would occasionally lift its head to look around, sniff the air with a twitch of its wet nose, then return to its siesta, its chin flat on the boulder.

Suddenly, the marmot bolted upright and let out an ear-splitting whistle that almost launched Sally from her high perch. It seemed incredible that an animal weighing less than ten kilograms could make such a loud noise. The marmot continued calling, staring straight ahead at the upper limit of the slope.

The warning call now echoed through the rockpile as other marmots repeated the distress signal. Each marmot stood at attention, alert and looking up the slope. I swung the camera in the same direction, but could not see what had upset the marmots.

Then a movement caught my eye. A coyote slipped from the shadow of a group of trees and began stalking a marmot, just fifty metres from us. I couldn't decide what to focus the camera on—the previously inert marmot that had come to life or the coyote on the ridge.

The coyote stood motionless, its coat burnished gold and copper in the sunlight. The animal leaned forward,

TOP: Sharing a rock with a marmot
BOTTOM: A marmot basking in the sun

Hoary Marmot

ears cocked and every muscle tensed for action. When the coyote took a few slinking steps towards the rockpile, all marmots dove underground except for one sentinel. Despite their short legs and heavy bodies, the marmots were able to break into a fast gallop and disappear in seconds.

The coyote made a desperate rush for the last visible marmot, but it too dove quickly underground. The alpine whistler's early warning system had, for this day, prevailed. They are not always so lucky: coyotes, mountain lions, hawks and eagles all take their share of this high-country rodent.

The call from the marmot we had been watching was now a nervous peep from under the rock; we could tell by its echo that it was descending to the depths of the rockpile. The frightened animal continued calling for almost half an hour. Gradually, the soft, short peeps grew louder as the animal returned to the surface. The marmot cautiously emerged from its den and stood on its hind legs, front paws tucked to its chest. The animal was silent now, and it was another ten minutes before it felt safe enough to venture further afield and continue feeding.

That action-packed episode was the last time we saw the marmots that summer. When we returned in mid-September it was already too cold for these creatures of comfort and they didn't emerge from their tunnels. They were hibernating in their deep, grass-lined dens, cuddled up in groups until the warm weather of next spring. To Sally and me, the alpine slopes seemed peculiarly quiet without the welcoming whistle of the furry creatures we had come to know so well.

See colour photographs on pages 74 and 75.

FIFTEEN

Feathered Friends

The air was heavy with cold when Sally and I headed out on a snowshoe trip from our remote wilderness cabin. The forest was still and the only sound was the soft swishing of snow from our snowshoes. We stopped occasionally and listened, anticipating the soft call of chickadees or the raucous laughter of grey jays who never failed to greet us as we passed through the woods.

The bold jays and captivating chickadees were feathered friends we could count on to entertain us in any season. These birds are true northerners. They remain all year long, not like the loons or geese who head for warmer haunts when the leaves turn colour.

I smiled when we finally detected the soft "tseee, tseee" of chickadees calling to each other as they moved in our direction. During winter black-capped chickadees assemble into loose flocks of eight to twelve birds and spend each day roving about the forest. They scatter quite widely, often so far they lose sight of one another but keep in contact by calling to each other, using their thin, lisp-

Chickadee and Jay

ing "tsee" note or the louder "dee-dee-dee."

Sally and I stood listening to their conversation and watching the flitting of wings in the tops of the spruce trees. The chickadees, with black caps and bibs and white cheeks, seemed cheerful even in the coldest weather. We marvelled at how these diminutive birds could survive in such low temperatures while we had to bundle up in many layers of clothing, topped off with Pak boots, parkas and heavy mitts.

"Chick-a-dee-dee," I called while digging into my jacket pocket for some of the grain I always carried with me. We waited patiently as the flock worked its way down from the treetops, each bird investigating the branches and cones in their never-ending search for food.

The chickadees flew with a quick flutter of wings, paused, then fluttered again in an undulating flight as they moved from tree to tree. These birds are airborne acrobats, able to land and take off from many different positions. They are able to alight upside-down on the undersides of swaying branches and hang there, apparently steady and secure. We noticed several birds doing this and were astonished to discover that they could take off from this awkward position, quickly righting themselves as they flew away.

I heard a fluttering in the spruce boughs beside me and turned my head slowly, not wanting to frighten the nervous bird which had settled on a branch within an arm's length.

"Pretty cold today, little fella," I said, as if talking to a friend.

The chickadee looked like a ball of feathers. The tiny creature had doubled its apparent size by fluffing its feathers to insulate its tiny body from the cold and to conserve valuable energy. I was amused to hear a clear "chick-a-dee-dee-dee" from this round bundle of fluff.

I held up my offering of grain and in an instant the chickadee was hovering a few centimetres above my

Chickadee and Jay

hand. It was so near that I heard the soft whirr of its wings, which flapped so quickly they were just a blur of grey. The chickadee did not quite dare to take the grain, however, and darted away in a moment.

Another bird also saw the food and flew nearer to study me. I must have looked quite harmless, for without hesitation it landed on my hand. Digging its tiny claws into my finger tips for balance, it leaned over and neatly picked up an oat flake. The chickadee was very light, weighing about as much as four pennies, and measuring only twelve centimetres from bill-tip to tail-tip. Even the oat flake looked large in its tiny beak.

The chickadee flew to a private branch to feed, then returned for more. After three oat flakes the chickadee had eaten its fill and began caching some in the trees. The bird stuck one flake under a piece of loose bark and the next in a patch of lichens. Then, as if not satisfied with the latter hiding place, the chickadee pulled the tidbit out and flew to another tree to repeat the tucking-away procedure.

More and more chickadees appeared, all eager to share in the feast, although some were more willing to risk reaching for the food than others. The chorus around us was now quite energetic. Their song was a short ditty, quite different from the "tsee" used to keep in contact when feeding. It consisted of two or three whistled notes with one higher, accented and prolonged note followed by one or two lower, shorter ones: "Fee-bee, feeee-bee-bee." A few sang their territorial call with a clear voice, repeating their own names over and over again.

Our other winter trail companions were the grey jays, or "whiskey-jacks," who often came by when we stopped for lunch. They are boisterous, grey-cloaked bandits who drop from the trees on folded wings to steal anything they can carry. These intelligent birds seem to know from experience that a human voice, the sound of an axe biting into wood or the smell of a campfire signals the possibility of food. Compared to the polite chickadees, the jays are ill-mannered and impudent, twice as large, and three times as loud.

We heard their raucous calls from somewhere in the forest; then, like a wisp of grey smoke, a jay floated into the clearing. A jay's usual method of travelling through the woods is to sail down from the top of one tree to the lower part of another, and then to hop upward from branch to branch, often in spiral fashion. They seldom flap their wings, except to gain altitude or cross an open area.

Our visitor glided downward on broad wings, making very little noise. The jay is a plump, long-tailed bird with a short, thick bill. It is cloaked in various shades of grey, from the nape of its neck to the tip of its tail, except for its throat, cheeks and crown, which are white.

The jay perched on a nearby branch. Like the chickadee, it sat with its feathers fluffed for insulation, making it look much larger than it really was. The jay cocked its head first to one side, then the other, studying us from different angles. Then, in a brazen display of nerve, it dropped onto my pack and probed inside for food.

Grey jays are opportunists and have been known to steal bacon from frying pans, food from tents, and even to take bites out of fish just landed by anglers. They will eat almost anything: seeds, the needles and buds of coniferous trees, insects, eggs and even carrion. What these birds can't eat immediately they coat with sticky saliva and paste under branches, tuck behind loose bark or stuff into cracks and crevices. The jays also make large stores in unused woodpecker holes, and always seem to remember where their food is stashed.

Soon after the first bird found us, several more arrived. The jays began to whistle imploringly, seeming to beg us to feed them. But their patience lasted only a moment, then one swooped down and landed on my hand, pecking at the sandwich I was eating. The jay's boldness was a

A grey jay eager to share our lunch

little disconcerting—if I hadn't had a strong grip on my lunch the bird would have flown away with it.

Another jay dropped on folded wings and hopped around at Sally's feet, calling plaintively with a soft "wheee-ooo." Sally was persuaded to toss the bird a substantial piece of her bannock. Three others immediately swooped down and began fighting over the prize, squawking and pecking, each stealing a bite before one jay claimed possession of the rest. That bird managed to corner the bannock and keep it away from the others by protecting it with outstretched wings. We watched it tear off piece after piece until its beak was overflowing with food. The jay grasped the rest in its claws and struggled to fly away with the heavy load.

During all this activity the jays were very vocal and loud, treating us to a surprising variety of sounds. Some were gruff and raucous, some shrill, and others full-toned and musical. The harshest call, with an overtone of aggressiveness, was their scolding note of "ack, ack, ack," which sounded like a short burst of toneless laughter. The softest, surprisingly, was the alarm note, reminding me of a quiet owl hoot, repeated twice.

As we shouldered our backpacks the jays sensed the feast was over and continued on their way. A few remained in the tops of spruce trees and performed a prolonged warbling song, punctuated with whistles and soft shrieks, a song so melodious it seemed they were thanking us for their meal.

Further down the trail we heard a flock of chickadees conversing as they flitted through the forest. I stopped and instinctively looked up at the trees. Even before I had spotted the birds, I reached into my pocket for a few of the morsels of grain that always seem to be lying there, caught in the seam for a visit from one of our feathered friends.

See colour photograph on page 68.

SIXTEEN
The Pine Marten

A low, ominous growl near the foot of my sleeping bag wakened me just before dawn. Sally and I were camping out under the stars and I peered into the semi-darkness, almost expecting to see the glint of fangs or sharp claws.

The throaty growl came again, this time even closer to where we lay. I could barely make out the outline of a dark animal as it slowly, stealthily approached across the snow. As my eyes adjusted to the dim light I finally realized what the fearsome animal was—just a pine marten, stalking us from two metres away.

The cat-sized animal kept its distance, circling around our sleeping bags, melting from position to position as it studied us. When I sat up to have a better look, the marten retreated with several long and easy bounds and climbed the nearest tree to perch on a swaying branch. The small animal uttered a soft purr-growl, then sprang to the trunk of the tree and spiralled upward in a blur.

Although martens are common in forested areas, until

The Pine Marten

now we had only caught fleeting glimpses of the animals. They are secretive and travel quickly and silently, either moving along the ground or leaping from tree to tree. But we had learned a great deal about martens' habits by following their characteristic paired footprints in the snow. One set of tracks we followed showed where a marten had meandered from tree to tree, checking out a squirrel midden and the resting place of a snowshoe hare before tunnelling into the snow.

We investigated one of the marten's abandoned tunnels by digging with a snowshoe and found a cozy den about thirty centimetres in diametre and lined for warmth with moose hair. Martens do not hibernate, so we found their tracks dotting the landscape in all but the coldest weather or worst snowstorms, when they denned up for a few days at a time.

The next time I encountered our neighbour was during a walk through the forest near our cabin. I heard the marten's familiar growl from a tree nearby, then a fan of spruce boughs parted and the marten timidly poked its head through. The marten's inquisitive, cream-coloured face featured black eyes and prominent black eyebrows.

The animal took a cautious step closer, its black nose describing an arc through the air as it sniffed for my scent. I could now see that the marten's throat and chest were splashed with pale orange and its thick winter coat of chestnut-brown blended with a buff-coloured underside. The luxuriously fluffy tail was almost the length of the marten's body and was draped loosely over the branch on which the marten crouched.

The marten examined me closely for a time, until its attention was diverted by the chatter of a squirrel. In a flash the marten whirled around and dashed up the tree in pursuit. With a high-speed display of arboreal acrobatics the marten and squirrel jumped from branch to branch, ascending higher and higher. The marten ran faster and leapt further than the squirrel. The distance

"Blackie" the marten prowling around camp

between them narrowed with each bound. The tension built as the squirrel's high-pitched chattering became a string of frightened churrs. This was a life-and-death situation: the marten was hungry and the squirrel would be a meal if caught.

In desperation, with the marten only a length behind, the squirrel scampered to the extreme tip of a branch. I thought the squirrel was doomed, but suddenly it leapt across a wide gap to another tree. Because the marten was heavier and could not venture out onto the slim branch, the chase was abruptly ended. The marten stared at the retreating squirrel for a moment, then descended the tree head-first, growling as though grumbling about its missed meal.

Although martens spend much of their time in trees, they catch most of their varied diet on the ground. Mice and voles are their most important food supply, but martens also prey on squirrels, chipmunks and hares. Martens will occasionally feed on frogs, beetles and grasshoppers, and they have a taste for birds as well—robbing nests on the ground and in trees. The energetic pace of this carnivore's life means it must constantly search for food.

The marten we had seen around our cabin began to visit frequently, checking out our bird feeder and looking for scraps of food. The marten was inquisitive, but always cautious, and vanished the instant we made any sudden movement or noise. We had learned not to make eye contact with the marten or any other wild animal because it may be taken as a sign of aggression. If we looked directly at the marten it would most likely bolt and disappear into the forest.

After many visits we decided that the half-metre-long marten was a male, and Sally named him "Blackie." He became bolder with each visit and began coming by at all hours. At night we could hear the marten pattering across the roof; during the day he often lounged in a pair of moose antlers that adorned the wall above the door. Whenever we opened the door Blackie would jump to the ground, then approach us with ears perked forward and head cocked to one side. He cautiously took any offered food, but usually stepped back a couple of paces to eat.

We looked forward to Blackie's visit each day and he probably looked forward to the treats we offered. He became more and more trusting until he was bold enough to eat scraps of food from our fingers and lick jam from our hands. His tongue was rough, but he was as gentle as a kitten, carefully grasping food between his needle-sharp teeth without biting our fingers. Blackie occasionally let us stroke him under his chin as he fed. He would purr softly, but if we reached up to pat his head he would immediately retreat.

One day after feeding Blackie I returned to my writing in the cabin, but left the door ajar. A few minutes later I noticed a movement out of the corner of my eye and looked up to see Blackie's small face peering around the corner of the door—two sparkling eyes, long whiskers and a wet, black nose.

"Come on in," I whispered encouragingly.

The marten slowly stepped into the cabin, padded a few steps, then paused while he looked and listened. The animal's movements were fluid and graceful and his flowing tail followed the undulating rhythm of his body.

Blackie's nose seemed to lead him around the cabin, sniffing at each new object he encountered. Sally drew in her breath sharply when he hopped onto the counter and peered at her sketches. She let out a gasp of dismay as the marten did a dance on her sketch pad, leaving muddy footprints across the white page. Next he pranced along the side counter and into the kitchen area. In a typically weasel-like pose, Blackie stood up on his rear legs and sniffed the aroma of fresh bread and other savoury delights.

Blackie dropped onto all fours again and followed his

nose to a corner of the cabin. What is this, a piece of cheese? He'd discovered a mousetrap! I quickly intervened by stamping my foot and the mischievous marten bounded out of the cabin.

A few days after Blackie's bold visit we noticed another marten bounding through the forest. Sally and I shared Blackie's excitement when the other marten appeared at the cabin one day in early March. This marten was considerably smaller than Blackie and was marked by a brighter orange patch on its chest. Blackie stood frozen in position, intently watching the intruder circle the cabin.

Suddenly, Blackie lunged at the other marten. They raced and shrieked through the undergrowth in such a blur that I could not tell one from the other. The sounds were as chilling as those of any cat fight, with snarls, growls and screeches. We became caught-up in the intensity of the fight, until the shrieks faded as the animals disappeared into the forest.

The next day we saw the two martens together, their differences apparently forgotten for the moment. It seemed that Blackie grudgingly shared his territory with the other marten; their tenuous friendship was regularly interrupted by snarling and hissing or wild chases through the trees. Even though the mating season for martens had passed months before, we assumed the visitor was female, because these animals will tolerate only members of the opposite sex in their territories.

In the spring the female disappeared for nearly a month. Because she had been a frequent visitor at our cabin, we began to worry that she had been caught by a lynx or fox. It felt unusually quiet around the cabin; even Blackie seemed to be less boisterous than usual.

Then one morning in early May we were wakened by a strange noise from under the floorboards near our bed.

"Meep... meep... meep."

Sally and I peered through the cracks between the floorboards, but could see nothing. We went outside to

The Pine Marten

circle the cabin and saw a fresh set of marten tracks leading to a hole under the foundation. Soon a familiar face poked out. As we were wondering why the female marten had tunnelled under the cabin, we heard the meeping sound again. Babies! That was why she had been away for a month.

The female must have decided that the space under our cabin would make a suitable den to raise her offspring: it was warm and dry, and the occasional morsel of food fell between the floorboards. It certainly seemed better than a marten's usual choice of an old woodpecker hole or squirrel den.

Although Blackie had tolerated the female in his territory, she would not allow him anywhere near her den. Whenever he came within sight she chased him with such aggression that he hastily retreated into the forest. I felt sorry for Blackie—he had shared his territory with the other marten and now he was forced to skulk around the cabin in order to visit us each day.

Within a few days the female proudly led two kits to the front of the cabin. The month-old kits were less than half the size of their mother and were lighter coloured than the adults. The young martens were clumsy creatures who tumbled and tussled like kittens. Their curiosity about the world around them seemed limitless, but they always kept one eye on their mother.

The mother marten took her babies on frequent outings into the forest, teaching them how to hunt. Once we noticed her, not far from our cabin, bounding along the ground after a chipmunk. The babies appeared to be eager students, although their clumsy movements accompanied by an occasional loud "meep" surely must have alerted any potential prey to their presence.

The martens often foraged around our cabin, prowling under the floorboards for mice or morsels of food. But when the mother began to spend more time at our cabin than teaching her young to hunt for food we realized that

Pine Marten

we were interfering with the ways of nature. We were corrupting the very animals we had come to admire. The next time the martens came by we chased them away, yelling loudly. It was difficult to turn away the animals we had made friends with, but we knew the young martens would have to learn to rely on their own hunting skills for future meals.

Although Blackie still hunted mice around our cabin, even his visits became fewer as he learned there were no more handouts. Every now and then we would see our neighbour peering out from the branches of a spruce—wild and free... where he belonged.

See colour photograph on page 77.

SEVENTEEN

King of the Mountain

High on the ridge a majestic, full-curl ram stood silhouetted against a panorama of snow-dusted peaks and blue sky. Through my binoculars the stocky animal looked as rugged as the crags on which it stood.

The bighorn sheep had spotted us long before we had seen him. He was looking our way and we knew that the strength of his eyesight far exceeded ours, even with eight-power binoculars. Knowing our every step was being watched, we made a show of descending until we were out of sight. Then we began climbing a gully where we could approach unseen. I tested the wind and gave Sally the thumbs-up sign—the valley had not yet warmed enough to create an updraft that would carry our scent flowing up the gully.

When Sally and I finally crested the ridge, crawling on hands and knees, we were only forty paces from a band of bighorn sheep. Several ewes and lambs rose from their resting places to have a better look at us. The ram we had seen earlier stood between us and the herd, as though on guard. Several other rams, who appeared to be smaller, were in a group a short distance from the others.

Each of the sheep had a dull brown coat of coarse hair, featuring a white muzzle, a white rump patch, and white stripes down the back of its legs. Their brownish-grey coats blended so well with the rocks that their large white rumps could easily be mistaken by a predator for an unmelted patch of snow. Each sheep had a stout body, fairly slim legs and an insignificant tail. The mature rams were crowned with massive curving horns, while the females had short thin ones that curved only slightly. Because the horns on the adult females were similar to those on the young males, it was difficult to tell them apart.

Unlike antlers, horns are not shed annually. By counting the deep growth rings on the horns we could guess the age of each animal and something of its life history. A wide space between rings meant food and conditions had been favourable that year, and more closely etched circles told of long, hard winters.

When we tried to move to a better position for photographs, the sheep walked away, keeping a safe distance from us. Even when the herd grazed on the alpine grasses it was never far from the safety of scree slopes or steep rock, where the sheep could easily outclimb us or a predator.

As long as we remained downhill from the sheep they were not too nervous, but when we climbed higher they spooked and bounded up a rock face. Sheep are accomplished mountaineers, being second only to mountain goats in the steepness of terrain they can climb. They bounded with short, stiff-legged leaps and seemed almost reckless as they careened from boulder to boulder. Then the sheep raced along a hint of ledge, where we would not dare venture without climbing gear. Moments later they disappeared from view over the top. No four-footed animal and few people would risk life and limb to follow them up and down their dangerous stairways.

We were left with nothing to photograph except the scenery. But we knew that bighorn sheep are creatures of habit and usually return to the same high meadow or ridge to feed, day after day. Well-worn trails and resting hollows scraped in the hillside confirmed that they frequented the area, and we were confident we would find them there the next day.

Sally and I arrived the following morning just in time to see two magnificent rams squaring off for a battle. They were almost equal in size, with impressive horns that began near their ears and grew back and around to form a complete circle. Each horn had a base thicker than my forearm and tapered to a narrow tip. The horns had grown so massive that the animals had scraped or "broomed" the ends against rocks to wear them down and keep them from blocking their vision. Even though the horns were heavy and appeared cumbersome, each ram carried his with an easy grace.

The sheep were so engrossed in their own drama that they didn't even look our way this time. The two rams paced back and forth, keeping their heads low and tipping their horns from side to side to show off their size and strength. Rams fight for dominance over others and thus earn the right to breed with the females. Although most fights occur during the rutting season of November and December, each animal must be ready to defend its position on the social ladder at any time.

The rams slowly approached each other, grunting and snorting. At two paces they stopped and stared, flicking their tongues in and out as a sign of aggression. One ram stepped forward and challenged the other by jabbing at his opponent's shoulder with a front hoof and making a series of abusive grunts, so expressive I could almost translate them into English!

Looking sideways at each other, they turned and walked ten paces in opposite directions. With perfect

A full-curl ram—king of the mountain

timing both rams turned simultaneously and rose onto their hind legs. They paused for a brief, suspended moment, then began a two-legged charge, churning rapidly towards one another, front legs awkwardly pawing the air. At the last moment they dropped onto all four legs for the final driving lunge. Just before collision the rams turned their heads slightly to the side. Their horns met with a resounding crash.

We were so awed by the spectacle that we simply stared, spellbound, forgetting to take any photographs. The combatants must have weighed well over a hundred kilograms apiece, and they collided with tremendous force. After the impact they stood motionless for almost ten seconds, appearing to be stunned.

There didn't seem to be a winner yet and we guessed, correctly, that the match was not over. The rams retraced their steps and rushed at each other again and again, their heavy horns meeting with a crack that made me shudder. The impact sounded like the thud of a heavy maul splitting wood and could be heard for many kilometres. If not for the sheeps' specially adapted skull structure and their impact-absorbing stance, these collisions would surely have broken their skulls.

Several other rams watched the confrontation with intense interest, but the ewes calmly grazed nearby, ignoring the contest. The battle continued until finally one male had endured enough head-bashing and staggered away. The dominant ram stood proudly, head held high. For now, he was king of the mountain.

The ram turned and stared down to where I was kneeling only a short distance away. In the way that only sheep's eyes can, the already large, amber eyes grew impossibly larger as he fixed his aggressive stare on me. The ram took a step forward. I stepped backwards. Like all hooved animals, sheep become unpredictable and aggressive during the rutting season; the last thing I wanted was a head-to-head confrontation with this animal!

Mountain Sheep

The ram must have been satisfied with my submissive retreat, because he turned and headed towards the females. He began herding them uphill, his upper lip curled back, hissing loudly to demonstrate his dominance. One ewe did not immediately rise and he kicked her with a front hoof to encourage her to join the group. The females meekly allowed themselves to be herded; they must have known that the contests and harrassment would go on for only another few weeks until the rutting season was over.

We photographed the herd for several days and watched the continuing drama as other rams challenged the leader. Most often, the challenge ended with a showdown but no battle. Any clashes that did occur were swift and decisive, and the ram was able to maintain his position of dominance.

Sally and I left the sheep a few days later when temperatures dropped and snow began to dust the rocky slopes, making the climb too treacherous for us. Within a few weeks, the sheep would also move down the mountain to their winter feeding areas. Because their small hooves are not well-suited for digging through snow, bighorn sheep generally migrate to lower ranges or choose open slopes, where wind and sun help expose the herbage they need to survive the winter. As the snow retreats in the spring, the sheep follow the snowline upwards to their summer ranges again.

We had no opportunity to visit the sheep that winter; our next chance to photograph them came in early June. We found them on the same mountain, although the band consisted of only ewes with young lambs and a few juveniles from previous years. The rams were nowhere to be seen and would not rejoin the ewes until rutting season in late fall. Now an old ewe led the group to feeding areas and away from danger. In fact, for most of the year a bighorn sheep herd is a matriarchal society, with the rams remaining in bachelor bands on separate ranges.

Mountain Sheep

The hillside echoed with the soft "baaaa-a-a" of ewes and lambs conversing with each other. The lambs were only a week or two old and remained close to their mothers, who would nurse them until the fall. One lamb began to suckle vigorously, its tiny tail wiggling in enthusiastic spurts of movement along with its impatient bunting. On their diet of rich milk the lambs had grown quickly. They were already able to keep up with their mothers as they grazed in the meadow.

The yearlings were independent and frolicked in small groups nearby. During one game a youngster leapt up onto a rock, only to be knocked off a moment later by another yearling who jumped up beside it. Often the game would change to a form of tag, with sheep chasing each other around and around, ricocheting from boulder to boulder at breakneck speed as they crossed the slope. They scrambled fearlessly among the crags, unconsciously gaining the mountaineering experience necessary for survival. All their dash and daring would be needed to evade the wolverines, wolves or cougars that occasionally prey on the sheep.

Later that day we watched two yearlings, each sporting short stubs for horns, engage in a mock battle. They butted heads, jockeyed for the uphill position and pushed at each other energetically. Already they were establishing their social position, with one male exerting dominance over the others. Through the years he would work his way up the social ladder until he attained his prime in seven or eight years' time. Perhaps this aggressive young sheep would then have his turn as king of the mountain.

See colour photograph on page 76.

EIGHTEEN

Smaller Friends

Sally and I awoke on a warm August morning to the pattering of tiny feet across the roof of our tent. The impudent creature scurried back and forth several times, its feet drumming on the canvas: up one sloping wall, across the roof and down the other side. For a creature that made a mere slip of a shadow on the roof, it certainly produced a lot of noise.

Quietly, I unzipped the door of the tent and peered out . . . peering back at me from the side wall of the tent was a chipmunk. The animal seemed frozen in position and stared with unblinking eyes. Then, with a flick of its long tail and a string of bird-like "chips," it scampered over the tent and disappeared into the forest.

Just as I turned back into the tent, another chipmunk emerged from nowhere and zoomed across the roof. Then another streak of tawny red scurried past only an arm's length away, its tail held vertically in typical chipmunk manner.

"This might be a good place for those chipmunk

Least Chipmunk

photographs we're looking for," I reported casually to Sally.

"That's a classic understatement," Sally laughed as she looked over my shoulder. The clearing around our camp was a hub of activity, criss-crossed by chipmunks with twitching tails and full cheek pouches!

As we were preparing breakfast I heard a rustling behind us, but thought nothing of it until I turned and spied a long, thin tail protruding from my backpack. Lured by the aroma of peanuts, raisins and other goodies, a chipmunk had climbed into the pack.

"Hey, get out of there!" I yelled, stamping my foot.

A small head emerged, cheek pouches stretched to capacity with booty, probably a tablespoon or more of my trail mix. The scavenger had the audacity to scamper only a short distance before burying its treasure in a shallow depression scraped in the duff. Another chipmunk arrived, looked me in the eye, then disappeared into the pack. Knowing from experience that there is no limit to the quantity a chipmunk will cache, Sally zipped the pack closed as soon as the second raider had departed. I was especially upset to see our precious supply of chocolate chips planted in the ground.

One of the chipmunks returned a moment later, probably expecting to fill its entire winter larder with the contents of my pack. I felt heartless as it scratched at the zipper with its tiny paws, eager to claim the food it could smell. Finally, the chipmunk chattered in disgust and dashed off.

When the chipmunks found they couldn't steal food from our packs, they changed their tactics and tried to charm it from us instead. They darted back and forth, coming a little closer each time, then sat on their haunches, churring quietly. Our determination quickly melted and we tossed a few tidbits to the ground. In a flash, the tiny creatures had snatched the treats and retreated to the safety of a willow shrub.

Least Chipmunk

Smitten by their antics, we tossed more granola, a little closer to us each time. When the crumbs were gone, one chipmunk became bold enough to scurry over my outstretched legs and brashly poke its nose into the pocket of my jacket, where I always keep birdseed. But as I reached to unzip the pocket further, the chipmunk was frightened away by my movement. I tried to coax it back by sitting quietly with my arm extended and hand resting on the ground. After a few minutes, the chipmunk returned and cautiously selected a few seeds from my hand before dashing away.

I was delighted when the chipmunk eventually became trusting enough to stay in my hand as it fed. The chipmunk was a warm, soft, quivering ball of fur which easily fit into the palm of my hand. I watched intently as it turned a sunflower seed over and over in its delicate paws, cracked open the shell with its sharp incisors, then stuffed the seed into a cheek pouch. Slowly, I raised my hand to eye level for a closer look at the friendly creature. This diminutive animal was a least chipmunk, the smallest of the chipmunk family. It was only twenty centimetres long—including its tail, which made up almost half that length.

The chipmunk's tawny coat was marked by five black and four white stripes running from nose to tail. The stripes are not just for the sake of appearance, but are a form of camouflage. This became clear to me when I accidentally frightened a chipmunk into a willow bush. I circled the bush, quite puzzled about where the animal could be, and finally spotted it lying stretched out and immobile on a branch where its stripes blended with the shadows. Chipmunks need this camouflage to help protect them from their numerous enemies, which include owls, hawks, weasels, martens, coyotes and foxes.

Sally and I spent many hours watching and photographing the chipmunks. While the chipmunks worked, they chittered and chirped to each other. They

Smaller Friends

used a steady "chip-chip-chip" when any large and potentially dangerous bird flew overhead. A softer "chit... chit... chit" seemed to be used for conversational chatter. This monotonous monologue could be kept up for half an hour, as chipmunks are able to converse even with their cheek pouches full of food.

Although the chipmunks seemed always to be dashing here and there, quivering with motion, we noticed that their activity was directly related to the weather. On bright, sunny days they dashed about the forest with vigour, but when the weather was cold or wet, a hush fell over their realm and they seldom emerged from their dens.

A few days later, as Sally and I were picking wild strawberries for a trailside snack, we saw a chipmunk streak by. The small creature plucked a berry with its tiny paws, then sat upright on its haunches with its long tail stretched out behind for support. To our surprise, it didn't eat the fruit of the berry, but painstakingly picked out the miniscule seeds, leaving a pile of pulp on the ground. Although they eat mostly seeds and nuts, chipmunks also eat mushrooms, insects and, on rare occasions, bird's eggs.

Oblivious of our presence, another chipmunk appeared and began feeding on grass seeds nearby. It stood on hind legs and reached up high with its dexterous paws to pull down a tall stalk of grass. The chipmunk nibbled the seeds from the cluster at the top of the stalk, stuffing both cheek pouches, then scampered off to its burrow in the forest. Most of the food a chipmunk collects is cached directly in its home burrow, but some is hidden in a number of caches nearby. The chipmunk's scientific name *Tamias*, which means "steward" or one who stores and looks after provisions, is certainly appropriate for these creatures.

Sally and I found out why it was necessary for chipmunks to have a number of caches when we watched one

A chipmunk balancing on willow branches

Smaller Friends

chipmunk make a food-stealing raid on the hoard of another. The thief entered the burrow, returning with bulging cheek pouches. It paused for a moment, perhaps looking for the cache's owner, then ran to its own burrow with the stolen food. The chipmunk returned shortly, repeating the robbery.

It wasn't long before the owner of the burrow returned and discovered what was happening. This time, as the thief emerged, he was pounced upon by the owner and chased away in a blur of stripes. The owner hurled a string of high-pitched chirps after the intruder, who escaped into the forest.

The resident chipmunk returned and immediately began to replenish the burrow from a supply cached nearby. We laughed, speculating that he, in turn, was borrowing from some other chipmunk's store. By hiding its food supply in a number of caches, known as "scatterhoards," the chipmunk had plenty of food set aside as a precaution against the inevitable raids on its main burrow.

Although we often saw several chipmunks in an area, each chipmunk has its own small territory. The territories often overlap, but each animal has a home burrow which it shares only during the short breeding season and while rearing young. In addition to home burrows, chipmunks have several others for escape from predators. Because their delicate paws are not adapted to digging, they choose soft soil or rotting logs in which to dig their modest homes.

After a few days of photographing the chipmunks, Sally and I left them to their business of survival. No doubt they had much work to do before fall, even though they are not true hibernators. Chipmunks don't put on a thick layer of fat, so they need an insulated home and a readily available supply of snacks under their nests for their periodic winter awakenings. We knew that by the end of October the chipmunks would descend into their burrows

Least Chipmunk

for the winter, plug the entrances and curl up for the long, cold season ahead.

Throughout our visit, we had been captivated by the charm of these little creatures. Whether dining on strawberries or scampering up our legs, the chipmunks reminded us that the most entertaining things in life often come wrapped in very small packages.

See colour photograph on page 70.

NINETEEN

Busy Beavers

The canoe slipped noiselessly through the water as we paddled up a quiet creek, enjoying the scenery and warm afternoon sun. Our ears were tuned to the subtle sounds around us: the trickling of water dripping off our paddles, the trilling of sparrows and, occasionally, the croaking of a frog in the marsh.

Suddenly, a loud "THWACK!" on the water shook us out of our reverie.

We jumped in surprise, almost overturning the canoe, and had hardly regained our balance when we noticed a movement in the water near us. A small black nose surfaced, followed by darting black eyes which peered about curiously.

Sally and I laughed about having been startled by the slap of a beaver's tail on water—after all, we were in beaver country. The large rodents had left abundant clues: along the shoreline were pointed, gnawed tree stumps and scatterings of fresh aspen and willow cuttings.

Beaver

We remained motionless, watching the beaver swim towards us, its ears perked up and nose quivering in the air. When it was about two paddle-lengths from the canoe the beaver must have caught our scent, because it slapped its broad, flat tail on the water again, splashing us as it dove. This loud tail-slapping warns other beavers of possible danger, and often scares a predator away as the beavers escape under water.

The beaver resurfaced and followed our canoe as we continued up the creek. It was a handsome creature, with glossy, dark brown fur. The beaver was more than a metre long from the tip of its nose to the end of its flat tail, and it was one of the largest beavers we had seen. We guessed that this beaver was the dominant male, checking for danger before the rest of his family came out of the lodge for the evening.

The beaver easily swam faster than we could canoe and soon passed us. We could see his front paws tucked close to his chest and his powerful, webbed hind feet alternating back and forth as they propelled him through the water. The beaver's paddle-shaped tail merely floated on the surface and was only occasionally employed to change his direction or to help him accelerate. There was a notch on one side of the tail, which I speculated might have resulted from an encounter with a hungry wolverine or wolf or lynx. Sally thought it could have been from a territorial dispute with another beaver. This old injury prompted us to nickname the beaver Notch-tail.

Around the next bend we found our passage obstructed by a long, curved beaver dam, which held back a deep pond. Like all beaver dams, the fifty-metre-long structure was an engineering marvel. Beavers construct dams by embedding sticks and twigs into the bottom of a stream at an angle against the current. The industrious animals pile on mud and stones, then add larger branches to build an incredibly strong structure.

Sally and I climbed onto the dam for a better view. The

still water mirrored a dome-shaped beaver lodge that thrust up from the centre of the pond. The two-metre-high mound resembled nothing more than a haphazard pile of sticks and mud, but was actually a carefully-designed, insulated shelter. Beaver lodges usually have two underwater entrances leading to a large living chamber. In this chamber is a platform just above water level where the beavers eat. The beavers also use this platform as a place to dry off, before entering the higher level where they sleep and raise their young.

Sally and I returned to the pond many times that summer. We usually arrived just as the sun was dipping behind the high mountains and could almost set our watches by the punctual appearance of the first beaver.

"Any time now," I whispered to Sally one evening as the shadows grew longer. Within a few minutes Notchtail emerged from the lodge and swam leisurely across the pond, cleaving the reflection of mountains and forest in two with his V-shaped wake. The beaver swam directly to the dam, then back and forth along its length, stopping periodically to inspect it for leaks. All must have been well with the dam, because he left without working on the structure and returned to the lodge.

A short time later a stream of bubbles heralded the arrival of another beaver. This beaver was followed by two youngsters, who immediately began chasing one another around the pond. They appeared to be playing hide-and-seek in the water playground. Suddenly, one youngster shot up in a splash of water, causing the other to slap its tail in surprise. Although these games seemed frivolous, we knew that they served a purpose: with each dip, splash and about-turn, the strength and reactions of the young animals were being developed.

The young beavers seemed more interested in playing than in assisting with dam building or tree felling. But the moment that one of the adults had dragged a tree to the water, both youngsters immediately swam over and

Dinner time for a beaver

began feeding. They were only six metres from us and we sat entranced, not daring to move a muscle. Even though mosquitoes buzzed around our heads and tickled our faces, we managed to remain motionless for a long time. When a voracious mosquito bit me behind my left ear, however, I flinched involuntarily. Notch-tail immediately slapped his tail on the water and every beaver disappeared underwater with a splash.

Sally and I groaned simultaneously. It had taken many days to get close to the beavers and one mosquito had jeopardized the feeling of trust we were trying to build. We remained at the pond until dark, hoping the beavers would return, but the wary animals did not resurface near our post that evening.

"Maybe we should specialize in mosquito photography," I said wryly. "There must be millions around here!"

It took several more visits before the beavers accepted our presence again. Each time Sally and I returned to the pond, we moved a little closer to the aspen grove where the beavers felled trees and fed.

We knew that the beavers had finally become used to us when Notch-tail waddled onto shore only three metres from where we sat. For the first time, we had a good look at him. Even though he appeared sleek and graceful when swimming, his rotund body looked cumbersome as he waddled awkwardly towards the trees.

After pausing to sniff and listen for signs of danger, Notch-tail went straight to an aspen. He stood up, leaning on the tree with his front paws and using his flat tail for support. Then he turned his head sideways and began tearing out chunks of wood with his large front teeth.

The work progressed quickly and within minutes there was a deep notch around the twenty-centimetre-thick aspen. He gnawed around the tree a second time, cutting deeper with each bite until the ground was littered with chips. The tree trembled on a tiny column. Then we heard a loud crack. Notch-tail put on an amazing burst of speed and galloped awkwardly back to the pond, out of the way of the falling tree.

Beavers are accomplished tree-fellers, although their round-the-tree method of cutting indicates they have no idea which way a tree will topple. This tree fell towards the pond, but beavers are not always so fortunate. We often came across abandoned trees that had fallen against other trees and become entangled there. Other aspens in the grove were only gnawed part way through, suggesting that the beavers had been interrupted by predators. For some reason, the beavers rarely seemed to return to finish cutting these abandoned trees.

Notch-tail waited in the pond for several minutes after the tree had crashed to the ground. Because beavers are at greatest risk from predators when they are on land, they spend as little time as possible away from the safety of the water. When he finally returned to shore, Notch-tail sniffed the air cautiously before waddling back to the aspen. He chewed off a large branch, grasped it in his teeth, then rushed to the pond, dragging the heavy branch behind him. The older beavers usually do the perilous job of felling trees, and only once did we see a young beaver come ashore to test its teeth on slim willow branches near the water.

As usual, the other beavers soon arrived to share the feast. Notch-tail didn't seem to mind at all. He severed a small branch for himself with two bites and held the branch in his front paws, turning it round and round with his deft "fingers." Bite by bite he nibbled off the bark with his chisel-like teeth. When the branch was completely de-barked, he discarded it and selected a leafy twig which he stuffed into his mouth.

After satisfying their hunger for the moment, Notch-tail and the two yearlings began to groom their fur. The beavers spent considerable time on this task; without constant grooming their fur would soon lose its water-

proofing and insulating qualities. Beavers use both front and hind paws to scratch and preen their hair, but the hind paws have a specially adapted split claw on the inside toe which serves as a comb. Every few minutes we noticed one of the beavers reach down to the base of its tail to extract oily castor from the gland there, and comb it into the fur. The beavers seemed to enjoy the grooming process, and often helped each other by combing backs and other hard-to-reach places.

A month after we first met the beaver family, they were joined by a pair of young kits, who followed their parents' example and showed little fear of our presence. The kits, born in late May or June, had remained in the lodge until they were old enough to swim. This family group of parents, yearlings, and kits would stay together until the following spring, when the older offspring would be forced out to find their own territories. This dispersal is the beaver's way of avoiding overpopulating an area.

One evening I cut some aspen branches and left them at the edge of the water near our observation place. The beavers' keen sense of smell must have alerted them to our offering of fresh bark, because they detoured and made a direct line to the branches. We noticed, after offering them several meals, that they invariably dined on the aspen first, leaving the birch for last. Many times the birch was left untouched. The beaver's favourite foods are aspen, alder and willow, although we had seen these beavers dine on grass and water plants as well.

As the leaves turned colour and colder nights came to the valley, the beavers began working more steadily, often coming out earlier in the afternoon and toiling through the night. It was time to prepare for winter, time to reinforce their dam and lodge with twigs and mud in order to withstand the rigours of the icebound season. The beavers also began collecting a supply of food by towing branches and twigs to the lodge and pushing them into the mud on the pond bottom. After a month of work, the cache of branches was almost as large as the lodge. Beavers do not hibernate, so this family would be making daily excursions under the winter ice to their food supply.

Sally and I returned to the pond on snowshoes in December, when the temperature was twenty below zero and the low sun remained hidden behind the mountains to the south. By now the only indication that beavers lived in the valley was a snow-covered mound on the flat pond. Crystals of hoarfrost at a breathing hole near the top of the lodge showed that all was well inside.

Many wolf and wolverine tracks led around and over the dome: those animals must have found the aroma of beavers enticing, even maddening. But the beavers were safe in their lodge. It seemed to us that beavers have a winter life without worry. Provided that their food lasts through the winter, they spend the season secure and comfortable under a sheet of ice and blanket of insulating snow.

See colour photograph on page 77.

TWENTY
Calling all Moose

The bull moose submerged his entire head in the lake, then came up with a splash, water cascading down his snout and plants streaming from his mouth. His massive antlers were unceremoniously draped with green strings of weeds and waterlilies. Hardly embarrassed about his undignified appearance, the bull plunged his head in again and again for a meal.

It was a summer afternoon and we had found the moose while canoeing across the shallow end of a northern lake. Because moose have few predators, and none which approach them through the water, the bull took little notice of us in the canoe, only twenty metres away.

The moose was a strange-looking creature. He had a large, overhanging snout and enormous ears, but very small eyes. A thirty-centimetre-long strip of tattered fur, called a dewlap, hung under his chin. When the bull moved closer to shore we had a better view of his deep-chested body, which seemed awkwardly balanced on thin, stilt-like legs. His massive shoulders and slim hind-

Calling all Moose

quarters appeared to be designed for two different creatures. The stubby tail looked like an afterthought and certainly would be of no help in swishing away flies; the only relief the moose had from bugs was to submerge his body in the water.

"Any closer and we'd better start backpaddling," Sally said nervously when the moose exhaled with a loud snort, sending a spray of water from his wide nostrils.

We had drifted to within two canoe-lengths of the bull and Sally was in the bow of the canoe, much closer than I was to the huge animal. Even though moose are relatively harmless and docile at this time of year, I agreed to move back to a more prudent distance—we didn't want to push our luck!

The moose also must have decided we were too close, because he waded deeper into the lake and began swimming to the other side. All we could see of the enormous animal moving across the water was a hairy snout, two long ears and large antlers. With each powerful stroke the moose surged forward. Moose are excellent swimmers and can paddle at speeds of up to eight kilometres an hour—far faster than Sally and I could ever canoe. Minutes later, the bull reached the opposite shore and paused to shake the water from his fur, flinging droplets high into the air. The moose casually looked at us over his shoulder then trotted into the bushes.

Sally and I returned to the marsh in late September, expecting to find the bull again. Each moose usually remains in a favourite marsh or valley bottom for much of the year, but the resident moose was nowhere to be seen. The rutting season occurs during September and October, so we came to the conclusion that the bull was busy searching for a mate. During this season, bulls lose all interest in eating. They also lose their quiet, mild-mannered disposition and become aggressive and unpredictable, ready to do battle with others for the opportunity to court a cow moose.

Moose

After searching the marsh for several days, we decided the only way we could photograph a bull in its prime would be to call one to us. Autumn is the time of year when bulls grunt challenges to each other, so Sally came up with a plan: she courageously volunteered to hide at the edge of the forest and take photographs... all I had to do was wade into the marsh and call a massive, temperamental bull to where I would be standing, knee-deep in the mud.

I waded into the marsh, armed with a moose-calling horn fashioned from a roll of birchbark. Then I raised the horn to my lips and coughed into it loudly.

"Gruuufff, gruuufff." This was the most convincing moose call I could muster.

I waited expectantly for a reply. After several minutes of silence I again picked up the birchbark megaphone and coughed another call. This time, the guttural grunt of a moose came from the forest.

"Gruuuuuuufffff," I replied with a long, drawn-out grunt.

There must have been a meaning to my call which the moose understood but I didn't. The effect was startling. From the forest came a hair-raising bellow, then something as big and as wild as a runaway horse came crashing through the bush. I heard the shearing of branches and shattering of small trees, punctuated by loud grunts from the moose.

As the bull crashed closer, a rapidly repeated "chock, chock, chock," like a string of curses, left no doubt in my mind that the moose was in ill temper.

"Maybe this wasn't such a good idea," I thought. I waited nervously, with second thoughts about standing knee-deep in the boot-sucking swamp. As I wondered just how big the noisy animal might be, it occured to me that the only place to hide was behind a clump of willows—a small clump of willows.

Then the head and front quarters of a huge, dark brown

moose emerged from the forest at the edge of the swamp. For several long seconds the bull stared at me with nostrils flaring, ears twitching and powerful neck muscles bulging. The animal stood taller at the shoulders than the largest saddle horse and had a palmated rack of antlers that spread two metres across. Bulls can weigh up to seven hundred kilograms, and this one must have been at the heavier end of the scale.

I felt very vulnerable and began waddling in my cumbersome hipwaders towards the dubious safety of the clump of willows. Only three steps later, I stepped into a chest-deep hole. I cursed under my breath as cold water flooded into the waders, then wondered how I would move with waders full of mud and water. I was in deep trouble if the bull decided that I was a competitor—running was out of the question, no matter how motivated I might be.

The huge moose stepped towards me, then tilted his head from side to side to display his massive antlers. I saw the hair on his shoulders rise up as stiff and straight as bristles on a brush. After a few minutes of this intimidating display the bull splashed across the marsh, his feet making a sucking noise with each step. He then shook a bush with his antlers and pawed at the ground, sending up a spray of water and mud. I remained absolutely silent—I had no desire to see what response another call or movement might evoke.

Then, as abruptly as he had arrived, the moose wheeled around and trotted off into the bush. I never did learn what prompted him to leave, but I was relieved by his sudden departure. As the moose entered the tangle of tall willow and scrub spruce, he tilted his head back and laid his antlers across his back so that they wouldn't catch or clatter on the trees. This time, the bull moved swiftly and silently, like a dark spectre gliding through the forest. I was amazed that such a huge animal could travel without a sound through the thick stand of trees where it

Calling all Moose

would seem a small deer could hardly pass.

"I got some great photos... call him back again!" Sally shouted from her hiding place as I emptied muck and swamp water from the hipwaders.

It took a while for Sally to persuade me to stake out a nearby marsh where we had seen a female on other occasions. If we were lucky, Sally said convincingly, we might even see a bull with the cow.

This time we didn't have to use the birchbark horn to call a bull because the female was doing her own seductive wailing. The cow called with a loud, whining voice and it wasn't long before a bull responded.

We heard crashing in the forest, and moments later a bull entered the clearing and trotted directly towards her. We could see that he had been called from his morning mudbath because his coat was thick with mud. During the rutting season bulls make a wallow by churning up a muddy depression with their hooves and antlers. Then they urinate in the muck and roll in the odorous mixture. This "eau de moose" seems to be agreeable to the females though, and they often roll in the mudbath made by the bull as a prelude to mating. I could imagine what this bull must have smelled like because the day before we had passed a wallow that was so rank it almost made us gag.

What followed could best be described as moose-courting, although it seemed anything but romantic from my viewpoint. When the bull was about fifteen metres away the female let out another whining "moooo-augh" that was obviously music to the bull's ears. Urged on by this enticing call he charged straight at the cow and stopped only a metre from her. The bull then followed the female around the marsh, grunting impatiently and shadowing her every movement.

The female was three-quarters the bull's size, had no antlers, and was of slighter build, especially in the shoulders. She had droopy lips and long ears like the

A cow moose only one canoe-length away!

bull, but her dewlap was considerably shorter. Although the cow had called the bull to her, she now paid little attention to him. She led the bull around the marsh for almost two hours before continuing on her way through the forest, with the bull still following close behind. Moose do not gather harems like elk or caribou; this bull would probably stay with the cow for a week or so before searching for another mate.

When Sally and I returned to the marsh several weeks later it was blanketed with new snow. The valley was criss-crossed with moose tracks, confirming that the moose were still in the area. These animals generally remain within a range of about ten square kilometres and migrate only a short distance to their winter feeding areas in willow flats or among protruding brush on the mountain slopes. During winter they forage for twigs. The name "moose" is derived from an Algonquin Indian word meaning "twig eater," a most appropriate description.

For much of the year moose are solitary creatures, but when the snow becomes deep, bulls, cows and calves often gather together in open areas where food is more plentiful. The constant trampling by many animals packs trails in the snow, making movement and feeding much easier. We often saw these trampled areas, known as "moose yards," during our winter travels. The moose would remain in each yard until the food was exhausted, then move on.

Not all moose gather in herds during the winter. One morning in early February we saw a cow and calf walking along the edge of a lake. We approached the pair from behind and downwind, but moose are especially wary during the winter because of predation by wolves. Our movements caught the cow's attention when we were still three hundred metres away.

I studied the cow through binoculars; she stood perfectly still, facing us for many minutes with her ears perked up and eyes alert while the calf nuzzled and nudged her impatiently. We were at a stalemate. It seemed that if we didn't move, neither would she. I slowly took a couple of steps closer, but the cow quickly turned away from us and began ploughing through the deep snow, lifting her front legs very high like a trotter. As she gained speed her hind legs lifted back and up above the snow, almost hitting her rump.

I could see only the head and shoulders of the small calf as it struggled through the trough left in the snow by the cow. We worried for a moment that the mother would keep going without the calf as the little one lagged further behind, but finally the cow stopped and waited for it to catch up. After a quick nuzzle of reassurance, the mother and offspring continued together into the safety of the forest.

Although the mother was protective now, we knew the pair would remain together only until the birth of another calf during late May. Then the yearling would be chased away to fend for itself. But after having watched moose trot through woods and marshes, swim across lakes and plough through deep snow, we knew the young moose would do just fine. Although they are strange-looking creatures, moose are well-suited for their lives in the rugged wilderness.

See colour photograph on page 72.

TWENTY-ONE

Through the Viewfinder

After many seasons of photographing wildlife Sally and I have learned that patience, knowledge of animals' habits, camera skills—and luck—all play a part in recording our furred and feathered neighbours on film.

One summer, for example, we headed into the mountains looking for goats and had climbed no higher than treeline when we spotted an Arctic ground squirrel basking in the sun. Without discussion, we changed our plans; we both unloaded our packs and prepared to spend the day photographing the furry creature instead of goats.

As we moved closer the squirrel scampered into its den, but I noted which hole it had entered.

"All we have to do is set up the camera at the den and wait until the squirrel peeks out," I said to Sally.

Just as I focused on the den I heard a loud "sik-sik" from behind us, and turned to see the squirrel at the entrance to another den only two paces away. I moved the tripod and camera closer and had just focused when the squirrel flicked its bushy tail and dove into the hole.

Arctic Ground Squirrel

Less than thirty seconds later the crafty creature popped up behind us, calling loudly from the first den. Again I moved the tripod and camera. Again the ground squirrel vanished just before I could focus. Matching wits with the ground squirrel proved to be a humbling experience. We thought we had finally solved the problem when Sally set up at one den and I stood ready at the other. But the squirrel peeked out from yet another hole, whistled as though to tease us, and dove underground again.

"Do you get the idea we're being outsmarted?" Sally asked, when several other ground squirrels joined in the game. The hill was busy with heads popping in and out of holes and Sally and me moving about the colony, trying to guess where the next animal might surface.

The odds were that sooner or later a squirrel would pop up at a hole near one of our cameras, but after more than an hour it became obvious that another plan was in order. We moved thirty paces away and hid behind a large boulder. Through the binoculars we studied the colony for several hours, watching the squirrels' antics and learning about their habits. We discovered that although they usually stayed near their dens, they would risk travelling far from the safety of their tunnels to feed on dandelions.

We began to move closer, giving the animals plenty of time to become used to our presence. After seeing them go to such effort for a dandelion, I thought that an offering of a flower just might lure a squirrel to within camera range. I placed a plump, juicy-looking dandelion at the entrance to a den and waited. A few minutes later a timid face peered out of the hole; the animal sniffed the air, then cautiously emerged, eyeing the gift flower. He'd probably never had room service before. As soon as the ground squirrel reached the dandelion, he pushed his nose into the middle of the bright flower.

After this close inspection the squirrel picked up the dandelion and sat on his haunches to enjoy this easy

A ground squirrel enticed by a dandelion

Arctic Ground Squirrel

meal. With both front paws gripping the stem, he shoved the entire flower into his small mouth. It was comical to watch the sunny mass of petals disappear bite by bite.

I looked up from the viewfinder and over to Sally. She was sitting cross-legged on the ground, bent over her sketch pad. The only sign of movement was her hand making slow pencil strokes across the paper. The squirrel Sally was sketching scampered closer, then began licking salt from the handle of her ice axe that lay nearby.

I smiled and thought back to our first days in the wilderness, when our enthusiastic but awkward advances resulted only in numerous photographs of the south ends of animals heading north, or even in inciting elk to charge us.

We had learned a great deal since then. Sally and I had gone to the wilderness as photographers; we returned as naturalists. We had learned to slow down, to observe, to wait. From the images recorded on film, to those recorded in our memories, our patience has been rewarded with many special encounters.

See colour photograph on page 78.

TWENTY-TWO

A Wilderness Legacy

When I remember our days in the wilderness I invariably recall the lonesome cry of loons on a lake, the bugling of elk in an autumn-bronzed meadow, or the scolding of a squirrel from a high tree.

The dictionary defines wilderness as "A wild region... a land with no people settled on it." But it is more than that. To Sally and me it is a place where we can drink water straight from a pristine mountain creek, where meadows of brilliant wildflowers create a kaleidoscope of colour, and where the smell of sun-warmed pine is the only scent on the breeze. Most important, it is a place where animals still live wild and free.

When Sally and I are in remote wilderness we are constantly aware of its irreplaceable value. No one can replace an old-growth forest after logging or a meadow of wildflowers once an open-pit mine has scarred the land; no one can return a river to its natural state once a dam has flooded a valley. Some areas need to be left as they are, kept free from development so that moose can

Conclusion

browse in quiet marshes, wolves can roam freely and other wildlife can live undisturbed.

It is our sincere hope that there will always be wild places where future generations will be able to find adventure and serenity. The wilderness is a legacy in trust, and time will judge if we abuse our obligations.

The high country—a wilderness legacy

REFERENCES

Our encounters with wild animals were always more rewarding when we knew something about the habits of the animals beforehand. Often we would pull out a reference book at the end of a day to learn more about the animal we had been with.

Below is a short list of guide books and reference books we found most useful in learning about our wild neighbours.

Burt, William Henry. *A Field Guide to the Mammals.* Boston: Houghton Mifflin Company, 1976.

Cahalane, Victor H. *Mammals of North America.* New York: The Macmillan Company, 1947.

Forsyth, Adrian. *Mammals of the Canadian Wild.* Scarborough: Camden House Publishing, 1985.

Gadd, Ben. *Handbook of the Canadian Rockies.* Jasper, Alberta: Corax Press, 1986.

Murie, Olaus J. *A Field Guide to Animal Tracks.* Boston: Houghton Mifflin Company, 1975.

National Geographic. *Guide to Birds of North America.* Washington: National Geographic Society, 1983.

Peterson, Roger Tory. *A Field Guide to the Birds.* Boston: Houghton Mifflin Company, 1941.

Savage, Arthur. *Wild Mammals of Western Canada.* Saskatoon: Western Producer Prairie Books, 1981.

Wooding, Frederick H. *Wild Mammals of Canada.* Toronto: McGraw-Hill Ryerson, 1982.

INDEX

Bear, black, 30-37, 66
   description, 30, 31, 32, 35
   diet, 30, 32, 34, 36, 49, 87
   habitat, 31, 34, 35
   habits, 30-37
   mating, 34
   voice, 31
   young, 34, 35, 37
Beaver, 77, 161-169
   description, 161, 162, 165, 168
   diet, 161, 165-169
   habitat, 161-169
   habits, 10, 161-169
   predators, 94, 162, 166, 169
   young, 163-168

Caribou, 45-52, 67
   description, 45, 46, 48, 49
   diet, 48, 49, 51, 52
   habitat, 49, 51
   habits, 45-52
   mating, 51
   predators, 45, 49, 94
   voice, 49
   young, 49, 51
Chickadee, black-capped, 68, 129-132
   description, 130, 132
   diet, 130
   habitat, 129, 130
   habits, 129, 130, 132
   voice, 129, 130, 132
Chipmunk, least, 70, 153-160
   description, 153-156, 159
   diet, 154, 157, 159
   habitat, 156, 159
   habits, 153-160
   mating, 159
   predators, 139, 143, 156, 157
   voice, 153, 156-159
Coyote, 75, 126, 128
   description, 126, 128
   diet, 109, 126, 156

Deer, mule, 73, 114-120
   description, 115, 117, 118
   diet, 115, 117
   habitat, 118
   habits, 114, 115, 117, 118, 120
   predators, 94, 117
   young, 118, 120

Eagle, bald, 61-64
   description, 61
   diet, 61, 62, 64
   habitat, 61
   habits, 61, 62, 64
   voice, 62
   young, 62
Elk, 11-20, 65, 66, 79
   description, 11, 12, 17, 19
   diet, 17

habitat, 17
habits, 11, 17
mating, 11-17
predators, 94
voice, 11, 12
young, 19

Fox, 101, 109, 142, 156
diet, 101, 109, 142, 156

Goat, mountain, 80, 81-89
description, 81-89
diet, 85
habitat, 81, 85, 87-89
habits, 84, 85, 87, 88
mating, 88, 89
predators, 87
voice, 85
young, 84, 85, 87
Goose, Canada, 69, 105-113
description, 105, 106, 110, 112
diet, 107, 112
habitat, 107, 109, 112, 113
habits, 105-113
mating, 107, 109
predators, 109
voice, 106, 110, 112, 113
young, 109, 110, 112
Ground squirrel, Arctic, 78, 179-182
description, 179
diet, 180, 182
habitat, 179, 180
habits, 179, 180, 182

predators, 94
voice, 179, 180
Grouse, ruffed, 38-44, 68
description, 39, 41, 44
diet, 44
habitat, 38, 39, 41, 44
habits, 38-44
mating, 38-42
predators, 41
voice, 44
young, 44

Harrier, northern, 58-61
description, 58, 61
diet, 59, 61
habitat, 61
habits, 58, 59, 61

Jay, grey, 132-135
description, 132, 133
diet, 133
habitat, 129
habits, 132, 133, 135
voice, 129, 133, 135

Lynx, 49, 101, 142, 162
diet 49, 101, 142, 162

Marmot, hoary, 74, 121-128
description, 121, 122, 126
diet, 122, 126
habitat, 121-123, 126
habits, 121-128
mating, 122
predators, 126, 128

189

voice, 121, 122, 126, 128
young, 125, 126
Marten, pine, 77, 136-144
  description, 136-142
  diet, 101, 137, 139, 143, 144, 156
  habitat, 136, 137, 139
  habits, 10, 136-144
  mating, 142, 143
  predators, 142
  voice, 136, 142, 143
  young, 143, 144
Moose, 72, 170-178
  description, 170-175
  diet, 170, 177
  habitat, 170-177
  habits, 10, 170-178
  mating, 171, 175, 177
  predators, 93, 94, 170, 177
  voice, 171, 172, 175
  young, 177, 178

Pika, 53-57, 71
  description, 53, 54
  diet, 54, 57
  habitat, 53
  habits, 53, 54, 56, 57
  predators, 57
  voice, 53, 56, 57
Porcupine, 21-29, 71
  description, 21, 22, 23
  diet, 22, 25, 26, 29
  habitat, 25, 26, 27
  habits, 21-23, 25, 26, 28, 29
  mating, 28, 29
  predators, 23
  voice, 21, 22, 26, 28
  young, 25

Sheep, mountain, 76, 145-152
  description, 145-150
  diet, 146, 147, 152
  habitat, 145-147, 150, 152
  habits, 145-152
  mating, 147, 149, 150
  predators, 152
  voice, 147, 150, 152
  young, 150, 152
Squirrel, red, 78, 98-104
  description, 98, 99
  diet, 99, 101, 102, 104
  habitat, 98, 99, 101
  habits, 98, 99, 101, 102, 104
  mating, 101
  predators, 101, 137, 139
  voice, 98, 101, 102, 104

Wolf, 90-97
  description, 94, 96
  diet, 45, 49, 87, 93, 94, 117, 152, 162, 169, 177
  habitat, 96, 97
  habits, 91, 93, 94, 96
  mating, 96
  voice, 90, 91, 97
  young, 96, 97
Wolverine, 49, 152, 162, 169
  diet, 49, 152, 162, 169

Also by Ian and Sally Wilson

"A unique, well written and exciting story..."
                                             CBC Toronto

This inspiring account of Ian and Sally Wilson's adventures in the wilds of northern Canada conveys the challenges and rewards of living close to nature—the accomplishment of building a log cabin 200 km from civilization, the intimacy of moose breath at three metres, and the isolation of a long, cold winter. Through each season Ian and Sally share their experiences in an area that is still wilderness—where moose, bear and wolf are as wild and free as the land.

A NATIONAL BESTSELLER
Published by Gordon Soules Book Publishers Ltd.
208 pages ISBN 0-919574-34-3 $12.95
Available at most book stores.